AA
Road Atlas of New Zealand

Published by Paul Hamlyn Limited,
Levien Building, St. Paul Street,
Auckland 1. New Zealand.
© Copyright Paul Hamlyn Limited 1974
Reprinted 1975
Printed in Hong Kong
ISBN 0 600 07347 5

GOVERNMENT HOUSE

FOREWORD

By His Excellency the Governor-General

Sir Denis Blundell, G.C.M.G., G.C.V.O., K.B.E.

     I congratulate the Automobile Associations of New Zealand on this splendid Atlas. Their initiative in preparing and publishing this continues a proud tradition of service to our motorists from the infancy of car travel early this century. Their wide experience is evident in the detailed maps, history and geography, and the valuable motoring advice to be found in these pages. Here indeed is something which combines practical value with wide interest.

     We have a deepening interest in places and buildings reflecting our Maori and pioneer heritage. There is ever increasing concern for the need for proper conservation. Our quite remarkable road system enables us to reach so many areas of scenic beauty. Yet too often much of this is missed in the bustle of fast travel and because people are unaware that these attractions are there to be found along their route or on short excursions from it. What is revealed in this Atlas will surely encourage many motorists, New Zealanders and visitors alike, to obtain greater enjoyment from their motoring and to learn so much more of the history and beauty of this country which we have in such full measure.

     I commend this Atlas to all motorists.

Denis Blundell

CONTENTS

Contributed by:*

St. John Ambulance.
Ministry of Transport.
Commission for the Environment

INTRODUCTION

The motor car, provides the means for New Zealander and tourist alike to enjoy the art of recreation.

New Zealand, with its beaches and rivers, mountains and plains, forests and grasslands, invites the wonderful pastime of exploring by car.

Motoring requires planning, preparation and thought. Day trip or holiday, alone or with the family, care must be taken to ensure that all goes well.

This book is aimed to assist you in all facets of your journey, whether business or holiday, spontaneous or planned. Use it to guide, educate, assist and possibly even save you.

THE AA ROAD ATLAS OF NEW ZEALAND is a reference and guide book for use in the car and at home.

The Automobile Association maps are the most accurate available and when supplemented with the AAs specific regional maps and booklets they give a complete guide to New Zealand.

The Atlas is divided into three main sections:

1. The Geographical Section consisting of regional and metropolitan maps, a most comprehensive gazetteer, plus historical and geographical notes.

2. The Roadside Identification Colour Section for guidance on your ventures around the countryside.

3. The Technical Section providing practical assistance for your journey.

Motor wisely.

Photograph · by Georg Kohlap

TROUBLE SHOOTING

The old saying "prevention is better than cure" applies to motor cars as much as it does to health. Proper preventative maintenance can reduce the risk of breakdown — have your car serviced regularly and preferably have it checked for wear and tear before undertaking any long journeys. If you would rather do this yourself, some points to watch are:

- **Radiator:** Check the radiator hoses and replace if they show signs of perishing. Make sure the radiator is full when the vehicle is cold. Add anti-freeze if you are going into snow country.
- **Battery:** Top the cells up to the correct level with distilled water.
- **Distributor:** Check points, cap, coil and all leads to make sure you are getting maximum efficiency from the electrical system.
- **Spark plugs** should be cleaned and the gap adjusted to the manufacturers' specification.
- **Oil:** Use the same grade and check the oil level every time you stop for petrol. Don't overfill.
- **Brakes:** Have them checked thoroughly and replace linings if necessary.
- **Tyres:** Check for bald or near-bald tyres. Fit new tyres if necessary and make sure the spare is fully inflated and in good condition. Have the steering inspected too.
- **Fan belt:** Make sure the fan belt, and generator belt if your car has a separate one, are correctly adjusted and in good condition. Replace if there is any sign of frayings.

No matter how thorough or comprehensive your precautions, breakdowns are a reality and the following advice may help you out of trouble.

Toolkit

Modern cars are usually supplied with a jack and a wheelbrace. These should be supplemented with:
- A torch
- A tow rope
- A set of combination spanners (check that they will fit your car).
- Pliers
- Medium screwdrivers
- An eight-inch crescent spanner
- Insulating tape
- Insulated wire
- A fan belt
- A well-sealed can of petrol
- Fuse wire or spare fuses
- Clean rags

Place all these in a clean sack suitable for lying on.

CAR WILL NOT START

Starter motor will not turn engine

Lights go dim

- Turn on headlights, engage starter.
- If lights go dim, check for loose or corroded battery terminals and leads.
- If these are corrosion free and tight, check for jammed starter. Some cars have a squared end on the armature under a thimble cap. Turn this with a crescent spanner to free the motor. If there is not squared end or slot for a screwdriver, engage an intermediate gear, make sure the ignition key is off and gently rock the car back and forth until the starter pinion disengages.

Starter motor will not turn engine

Lights remain bright

- Check for a loose or broken key start wire.
- If these are intact and making proper contact, the trouble could be a faulty solenoid. Many English cars have a push button in the centre of the solenoid. Ensure that the key is on and the gear lever in NEUTRAL BEFORE pressing the button. For cars without a solenoid it is possible to BRIDGE the solenoid terminals with a heavy screwdriver, but this is recommended only in EXTREME emergencies.

Starter turns motor

- Make sure there is sufficient fuel in the tank FIRST.
- Remove the air cleaner and open the throttle. Fuel should spurt from the jet. If it does not, check for a blockage or air leaks at the fuel pump or in the pipes back to the tank.
- If there is fuel at the carburettor, check for spark. Remove a high tension lead from a spark plug and hold it about 5 mm (¼in) from any metal part of the motor. With the key on, engage the starter and watch for a healthy blue spark between lead and metal.
- If there is no spark, check for loose or broken low tension connections at the coil and distributor.
- Remove the distributor cap and make sure that the rotor turns and that the contact breaker points are clean and opening and closing when the motor is turned over.
- If none of these measures work, it may mean that the motor is flooded. Push the accelerator gently all the way to the floor and engage the starter. Do NOT pump the accelerator.

PUNCTURES

Make sure the spare is not flat too before you start to change the puncture. Place it beside the wheel with the flat tyre. If the car is stopped on a slope, place a rock, piece of wood or some other suitable object behind a wheel on the opposite side of the car. This ensures that the car will not slip off the jack.

Before placing the jack in position as shown in the manufacturer's handbook, loosen the wheel nuts. The brace supplied with the car may have insufficient leverage to slacken wheel nuts tightened by a garage. Use the extra tools as shown in the illustration. If extra levering implements are not available, jack up the car, place bricks are something similar under the wheel brace as shown in the illustration and lower the car.

Ensure that the bricks are to the right of the nut.

Once the nuts are loosened, jack up the vehicle and change the wheel.

Before driving off, make sure that the wheel nuts are as tight as possible and that the hub cap is firmly in position.

TYRE BLOW OUT

A blow out need not lead to loss of control of the car provided that you keep both hands on the steering wheel and DO NOT PANIC. If the blow out occurs on the right-hand lane of the motorway and the traffic is heavy, it is better to run on to the grass median strip once the car is under control than to attempt to cross the hard shoulder.

Rear Tyre Blow Out

The rear of the car will bump and sway, but if a firm grip is kept on the steering wheel the car can usually be held on a reasonably straight course. Pump the brakes with an on-off movement. This throws the weight of the car on to the front tyres and relieves the strain on the blown tyre. Do NOT pump the brake harder than is necessary.

Front Tyre Blow Out

This will seriously affect the steering, so grip the steering wheel firmly with both hands to counteract any violent changes in direction. Brake gently and avoid throwing the weight of the car on to the front tyre.

BURST RADIATOR HOSE

The first indications of a burst radiator hose are engine misfiring caused by water spraying on to the ignition and then excessive pinking.

Stop as soon as practicable and allow the motor to cool down. Do NOT add cold water to an over-heated radiator as this can cause serious damage.

Bind up the split with insulation tape.

Fill the radiator with water. If no houses are nearby, look for a stream or use water from a thermos flask or screen-wash bottle. To prevent pressurisation of the cooling system which may force water out through the split, leave the radiator cap loose and drive gently to the nearest garage.

BROKEN FAN BELT

The first signs of a broken fan belt are continuous illumination of the ignition warning light, followed by overheating.

If no spare is available, pantyhose or a nylon stocking can be used as a makeshift fan belt. Slacken off the generator adjustment and tie the stocking or pantyhose around the pulleys as tightly as possible. Re-tension the substitute, turn off all unnecessary accessories and drive quietly to the nearest garage. Check that the radiator contains sufficient water, but do NOT add cold water while the radiator is still overheated.

LEAKING RADIATOR

The first signs of a leaking radiator are continuous illumination of the water-temperature warning light followed by engine misfiring.

Allow the engine to cool down and refill the radiator with cold water.

To prevent pressurisation of the cooling system which will accelerate water loss, leave the radiator cap loose and drive quietly to the nearest garage.

In an extreme emergency, a handful of oatmeal, a raw egg or a tablespoon of mustard will seal a leak on a TEMPORARY basis. The permanent repair must include complete flushing of the cooling system.

BLOWN FUSE

Failure of any of the electrical accessories such as windscreen wipers, headlights or horn may be caused by a blown fuse. The fuse box is usually located underneath the dashboard or in the engine compartment on the bulkhead or the wing. The handbook supplied with the car should indicate its position.

If no replacement fuses are available, ordinary household fuse wire can make a temporary repair using the original fuse casing as a support. ENSURE that the current value is the same as that for the original fuse. Alternatively borrow a fuse from a non-essential accessory such as the radio.

Do NOT use silver paper as this may lead to a serious fire.

CLUTCH SLIP

One of the first signs of clutch slip is a rise in engine speed without a consequent rise in road speed, particularly on hills. Ensure that the clutch mechanism has the correct clearance and drive to the nearest garage. Avoid gear changes as much as possible.

TROUBLE TRACING

Battery flat

Lead disconnected
or corrected

**STARTER
DOES NOT
TURN
ENGINE**

**STARTER
TURNS
ENGINE
SLOWLY**

Battery partly run down,

Connections dirty,

Terminal loose.

Faulty starter switch

Faulty starter motor

Faulty starter motor.

**STARTER
TURNS
ENGINE
FAST**

IGNITION
**NO SPARK
AT PLUG
LEADS**

**SPARK AT
PLUG LEADS**

Spark plugs oiled up,

Spark plug cracked,

Plug leads disconnected.

DISTRIBUTOR

Cracked rotor,

Loose low tension lead,

Faulty distributor cap,

Dirty distributor cap,

Dirty distributor points,

Faulty condenser,

Carbon brush not making contact.

FUEL
**FUEL
SYSTEM**

**IF NO PETROL
IN CARBURETTOR**
Air leak in fuel line

Blockage in fuel line

Faulty petrol pump.

**PETROL AT
CARBURETTOR**

Jets choked,

Carburettor mounting loose,

Water in petrol,

Dirt in carburettor.

COIL
**NO SPARK
AT COIL
LEAD**

Burnt out coil,

Loose or broken high tension lead,

Faulty switch

BATTERY

ENGINE MISFIRES

High tension leads shorting,

Incorrect spark plug gap,

Cracked spark plug porcelain,

Battery connections loose,

Electrical lead damp.

Water in carburettor

Petrol line partly blocked,

Faulty fuel pump,

Fuel pump filter choked,

Needle valve faulty or dirty.

Sticking valve,

Broken valve spring,

Incorrect valve clearance.

ENGINE STARTS AND STOPS

Low tension connection loose,

Faulty switch contact,

Dirty contact points.

Petrol line blocked,

Water in petrol,

Needle valve sticking,

Fuel pump faulty,

Air leaks.

ENGINE SLUGGISH

Petrol feed faulty,

Ignition timing incorrect,

Carburettor incorrectly adjusted.

Valve sticking,

Valve burnt of broken

Valve spring broken.

ENGINE DOES NOT GIVE FULL POWER

Ignition retarded,

High tension lead shorting,

Faulty distributor cap,

Faulty plug.

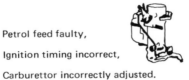

Valve burnt or bad seating,

Incorrect valve clearance,

Petrol supply faulty,

Jets partly choked.

ENGINE RUNS ON WIDE THROTTLE ONLY

Slow running jet blocked,

Carburettor flooding.

ENGINE KNOCKS

Timing over advanced

Excessive carbon.

Worn bearing or piston.

KEYS LOCKED INSIDE THE CAR

Look for an unlocked or insecure quarter light first.

If entry cannot be gained this way, form the end of a piece of wire into a semi-circle. Thread this around a window, usually between the window frame and the weather strip, and hook up the internal lock button or door catch.

CLUTCH FLUID LOSS

Early indications are a spongy pedal and difficulty in engaging and changing gear.

If the reservoir is empty and the leak is due to cylinder seal failure, it is permissable in extreme emergencies to fill and bleed the reservoir with water then drive to the nearest garage for proper repairs.

BRAKE FLUID LOSS

The first signs are a spongy pedal followed by eventual loss of brakes.

If the reservoir is empty is it advisable to have the car towed to the nearest garage for proper repairs.

If a tow wagon is not available, the car may be driven very slowly to the garage using the gears and handbrake.

If another car is available, use the brakeless car to tow the other vehicle, which can then do the braking for both.

Only in extreme emergency, when there are no possible alternatives, may water be used to top up the reservoir. The car must then be repaired properly at the nearest garage.

STUCK IN MUD

If one wheel is spinning and the other is locked, apply the handbrake gently. This will tend to lock the differential and transfer some of the drive to the stationary wheel. Do not spin the driving wheels down until they make a trough, but engage low and reverse gears alternately and attempt to rock the car on to firmer ground.

NEGOTIATING A FLOODED ROAD

Try to find out the depth of the water. If it is more than 300 mm (10-12ins) it is advisable to remove the fan belt, engage a low gear and drive slowly through the water. Keep the motor running fast as this will prevent stalling and water from entering the exhaust pipe.

Once clear of the water, refit the belt and dry out the brakes by applying them until, they are even and effective before regaining normal speed.

A large proportion of A.A. Breakdown Service calls, are to people with "drowned motors", particularly the cars with an east west motor, with the distributor situated in front of the vehicle.

Water is an excellent conductor of electricity, and if a motor dies, after or while negotiating a flooded area, it will be necessary to "dry out the electrics". There are many excellent "de watering fluids", contained in an aerosol type can on the market, and these can be used in this contingency. However, if you do not have one of these, take a piece of dry clean rag and clean off the moisture laden dirt from the high tension leads, the plug insulators, the distributor cap, both inside and out, and the coil tower. It may be necessary to remove the high tension leads for thorough drying. Remove and replace these one at a time, or you may lose the correct firing sequence.

It is worth noting that there are water repellent silicone based preparations, also contained in aerosol cans that can be sprayed onto the high tension components, and these reduce the possibility of moisture causing a breakdown in the future.

ESCAPING FROM A SINKING CAR

If you have the misfortune to drive into very deep water — for instance off the end of a jetty — do your best to remain calm.

Some cars will float for a short time. If yours does, get everybody out as quickly as possible through the windows.

If the car sinks quickly, close all the windows immediately, release the seat belts and get the heads of children and any injured passengers up into the remaining air pocket. If there is time, turn on the lights to aid rescuers.

When the water reaches chin level try to open the door as the water pressure inside and outside the car will by then be approximately equal. Form a human chain and float to the surface. This will ensure that nobody remains trapped in the car and that assistance can be given to each other on the surface.

BROKEN WINDSCREEN

If your car has a ZONE-toughened windscreen which is shattered by a stone or similar object, a reasonably clear patch of screen will be left in front of the driver. This will allow the car to be driven carefully to a windscreen specialist.

Ordinary toughened glass shatters into an opaque crystalline structure which makes forward vision impossible. If this happens while the car is being driven, do not panic. Look out the side windows and follow the original direction in which the car was moving. Slow down, check your rear vision mirror and pull off the road when it is safe to do so.

If forward vision is reduced to a dangerous level it is advisable to remove a section in front of the driver. This is best done as follows:
- Cover the plenum-chamber and windscreen-demister vents with pieces of rag.
- Wrap a cloth round your hand and push OUTWARDS a section of windscreen in front of the driver.
- Clear away all fragments of broken glass. Wear sunglasses if available and drive carefully to a windscreen specialist.

STEERING

If the car wanders, the first check would be the tyre pressures, they must be to the manufacturers specifications. Add a further 2lbs per square inch if the tyres are radials. Make sure that the tyres are a pair, both in size and type, particularly on the front. Ensure that the car is not overloaded, check that the roof rack, if fitted, is not overloaded. This alters the centre of gravity and can be affected by side winds. Check for uneven or broken springs, play at the steering box or joints and ensure that the steering geometry is to specifications.

If the steering is heavy, correct the tyre pressures. Add an extra 2lbs per square inch if the tyres are radials. Ensure that all moving parts of the steering are lubricated and that the steering box is correctly adjusted. Adjust steering geometry to manufacturers specification.

If the wheels shimmy, it is practically certain to be an imbalance of the front wheel and tyre assemblies. Other causes are buckled wheels, out of round tyres, incorrect steering geometry, worn steering components, and incorrect front wheel bearing adjustment.

If the car pulls to one side, check that the tyre pressures are correct and also that the tyres are a pair, both in size and type, particularly the fronts. Other causes are seized, bent, worn or insufficiently lubricated steering components, misaligned chassis, weak or broken springs, misaligned or loose rear axle and incorrect steering geometry.

Rapid tyre wear may be caused by over enthusiastic driving, such as hard cornering, fierce acceleration, or fierce braking. Other causes, are incorrect tyre pressures, unsuitable tyres for that particular application, and incorrect steering geometry.

BRAKES

If the brakes pull to one side, suspect oil or brake fluid on one or more brake linings, seized wheel cylinder, incorrect brake or wheel bearing adjustment.

If the brake pedal has a long travel but feels firm, the most likely cause is too much clearance between the brake shoe and brake drum. Adjust the brake shoes, but ensure that there is plenty of brake lining left on the shoes.

If the brake pedal has a long travel and feels spongy suspect air in the hydraulic brake system. Bleed the brakes with new brake fluid and adjust. Get somebody to stand hard on the brake pedal and carefully inspect the entire brake system for fluid loss.

If the brake pedal goes straight to the floor, check the brake fluid level. If there is sufficient fluid in the reservoir, suspect that the master cylinder piston seal has collapsed. If the brake reservoir is empty, check for leaks at the master cylinder, brake hoses, pipes or wheel cylinders.

If the brakes grab, a likely cause will be a contaminant on the brake linings, such as grease from a leaky wheel bearing seal. Other causes are incorrect brake lining material, loose anchor pin or loose backing plate.

If the brake pedal pulsates on application, suspect an out of round brake drum, poorly fitted brake linings or wheel bearings out of adjustment.

If the brakes drag or bind, they may be overadjusted, or the brake master cylinder push rod may have insufficient clearance, or the brake master cylinder return port could be blocked by foreign matter or a swollen piston seal.

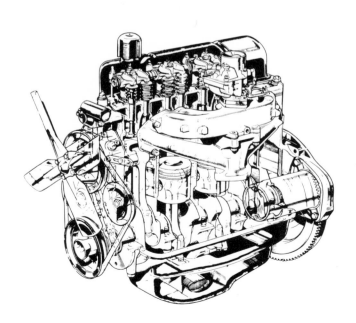

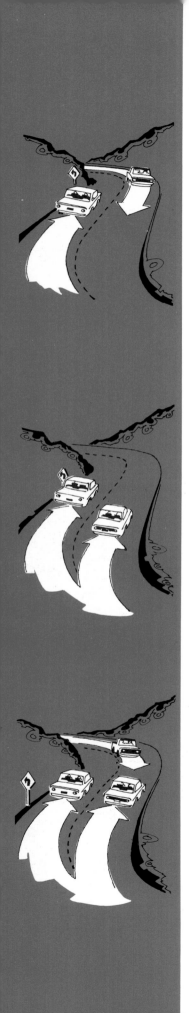

ROAD CODE

A driver's licence does not automatically make you a good driver. Competent driving is a skill and requires complete concentration whenever you are behind the wheel. There are certain habits which are the hallmark of a good driver; among the more important are:

- **Punctuality.** Allow plenty of time for any journey. Haste can lead to carelessness and the taking of unnecessary risks.
- **A sense of fair play.** Remember that the roads were not built solely for your benefit and that the road code is intended to do more than interfere with your freedom.
- **Modesty.** Do not try to impress your passengers with your fast driving or try to prove to another motorist that you are a quicker driver. It is even more dangerous to attempt to demonstrate that liquor has no effect on your driving.
- **Concentration.** The car is potentially a lethal weapon and should always be treated with the respect it requires. Keep your undivided attention on your driving and do not turn around to talk to passengers or to admire the scenery.
- **Intelligent Anticipation.** Be prepared for unexpected events: a truck that slows to turn off or pedestrians, especially children, who suddenly cross the road.

The competent driver also has a thorough knowledge of the road code too. The following notes have been prepared as a guide by the Ministry of Transport. You may be surprised at the number of things you have forgotten. It pays to re-read the Road Code occasionally to refresh your memory.

KEEP LEFT

Keep as close to the left of the road as practicable. Don't cross the centre lines at bends, near the crest of a rise or anywhere else where you can't see clearly for at least 100 yards (90m) ahead. Cutting corners is dangerous — don't do it.

OVERTAKING

You must overtake on the right, except in a few special cases:
- If you are directed by a traffic officer
- Where there are two or more lanes on your side of the centre line and you can pass safely and considerately.

Before passing another vehicle, make sure that the view ahead allows you to pass and still have at least 100 yards (90m) of visibility left when you get back into line. Check behind for other traffic and give a pulling-out signal. After overtaking, get well back into line. And when you are being overtaken, never speed up to prevent another driver passing.

It is illegal to overtake under all circumstances when:
- A vehicle has slowed or stopped to give way to pedestrians on a pedestrian crossing, or a train at a railway crossing, or at an intersection (except in special circumstances mentioned later)
- Within 30 feet (9m) of a railing crossing
- When approaching a blind bend, the crest of a hill, or within 30 feet (9m) of an intersection (again, except in special conditions outlined below).

NO PASSING LINES

These are indicated by a solid yellow line on your side of the broken centre line or by a double solid line.

You may not overtake at "no passing" lines unless you keep entirely to the left of them, and there are two or more lanes marked on your side of the roadway, or you have at least 100 yards (90m) of clear visibility throughout the whole overtaking movement.

OVERTAKING AT INTERSECTIONS

You may overtake at intersections under certain conditions, but it is your responsibility to ensure that the movement can be carried out safely.

Where lanes are marked, you may overtake if the vehicle being passed is in its correct lane to go straight through or has indicated a right turn. You may also overtake on the right when approaching an intersection, if the vehicle you are overtaking is not in a lane from which a right hand turn is permitted and as long as you don't cross over the centre line.

Where lanes are not marked outside a 50 km/h area, you may overtake on the left when approaching an intersection if the vehicle you are passing is stationary or has signalled a right turn. You can pass too when the vehicle being overtaken is as far as possible to the left and you don't cross over the centre line.

When leaving an intersection in a 50 km/h area or in any area where lanes are marked, you may overtake on the left if the vehicle being overtaken has just made a right turn or has gone straight through.

You mustn't overtake any vehicle in any area when it has slowed or stopped to allow a pedestrian to cross on a pedestrian crossing.

INTERSECTIONS

An intersection is defined as any road or street junction or crossroads, regardless of the number of roadways or angles at which they join. You must not enter an intersection when your passage across it, or your exit from it, is blocked by stationary traffic.

At intersections where there are lanes, you must obey the directions indicated by arrows or words marked in them. You mustn't change lanes so suddenly that you inconvenience or endanger other traffic.

When turning left, make your turn from the extreme left of the roadway or use a left-turn lane where provided.

Right turns are more complicated. In 50 km/h areas where it is safe, move just to the left of the centre of the roadway and, when the way is clear, turn as directly as you can. Move back to the left as soon as it can be done safely.

In other areas, turn directly to the right if the road is clear.

If this could endanger or inconvenience other traffic, pull over to the left (on to the shoulder if there is one) and wait for a gap in the traffic.

Where there are lanes marked, use the right-hand lane or any lane marked for a right turn. In turning, slow down and signal your intention well before you reach your turning point. You must not increase your speed at an intersection which any other vehicle is approaching or crossing, whether you have the right of way or not.

RIGHT-HAND RULE

Where there are no traffic lights, pointsmen, Give Way or Stop signs, give way to **all** traffic on your right. When turning right give way to all other traffic.

If two vehicles are turning right, the right-hand rule is superseded by the law of courtesy.

ACCIDENTS

If you have an accident, you are legally obliged to stop immediately and help any injured person in whatever way you can.

Accidents in which injuries are received must be reported to the police by the driver as soon as possible within the following 24 hours. If no-one is injured, you must give your name and address to the owner or driver or any other vehicle which has been damaged and also to the owner of any property which has been damaged. If you cannot find out the name of the driver or owner, you must report the accident to a traffic officer or the police within 48 hours. Also, the owner must notify the insurance company covering his third-party risk.

The same type of procedure is involved when a driver injures an animal. The accident must be reported to the owner or person in charge of it. If this person cannot be found, report the accident to the police or a traffic officer.

WET-WEATHER DRIVING

Extra care is needed in adverse conditions. This includes making sure your car is safe by checking the brakes, tyres, steering, windscreen wipers and lights.

The windscreen must be clean and clear. In fog, dip your headlights so that you'll see and be seen more easily.

When beginning a trip in wet weather, you need to get the "feel" of the road. Wet surfaces give less grip to tyres which means that braking becomes more difficult. You can prove this by checking to see that there is no following traffic then cautiously trying your brakes.

Because it takes longer to stop on wet roads, you need to allow extra (at least double) following distances. When you brake, pump the pedal gently. Jamming the brakes on suddenly could throw the car into a skid.

If you are involved in a skid, do not jam your foot on the brake or push in the clutch. This will only make matters worse. Instead, take your foot off the accelerator or brake (whichever is being depressed) and turn the front wheels in the direction of the skid. As the car straightens out, straighten your front wheels.

NIGHT DRIVING

You must display your headlights during the legal hours of darkness. These are defined as 30 minutes after sunset and 30 minutes before sunrise and any other time when there is insufficient daylight to clearly see a person or vehicle 50 yards (45m) away.

When driving during darkness, you must limit your speed so that you can stop well within the ranges of your headlights. Your lights should be dipped for oncoming traffic, when following another vehicle, when street lighting gives good visibility or when you stop. Also dip your lights when approaching a pointsman.

MOTORWAYS

There are two rules relating specifically to traffic on motorways:
- You may not walk, cycle or ride on a motorway.
- You must not stop your vehicle.

Your driving will also need to be modified. Making "U" turns is illegal, for example. Signal well before changing lanes and if you are driving slowly move to the left lane. Exit lanes are well signposted, and it is wise to get into the correct lane early.

If your vehicle happens to break down, get it as far off the carriageway as possible. Hang a white cloth from the driver's door handle or window and raise the bonnet if possible. It is best to remain in the vehicle if it is clear of the carriageway, but if you must walk for help, keep well clear of the traffic lanes.

PARKING AND STOPPING

Rules about parking are designed to ensure that parked cars block the road as little as possible and do not inconvenience drivers and pedestrians.

You must always park parallel to the roadway and as far as practicable to the left, unless angle parking is indicated by a sign or road markings. It is illegal to park on the right-hand side except in a one-way street where you may park on both sides unless signs indicate a restriction.

Other areas where you may not stop, stand or park any vehicle are:

- Near any corner, bend, rise, safety zone or intersection where your vehicle might obstruct other traffic or another driver's view of the road.
- On the roadway when it is possible to park clear of it without damaging grass plots or cultivated frontages.
- On any footpath.
- On an intersection or within 20 feet (6m) of one.
- On a pedestrian crossing or within 20 feet (6m) of one.
- Opposite a safety zone
- On a marked bus stop or within 20 feet (6m) of one which is indicated by a sign on a post.
- In front of a vehicle entrance.
- Alongside another parked vehicle.
- On yellow "no stopping" lines.

MOTORISTS AND PEDESTRIANS

There are times when pedestrians take precedence over motorists. Special care is needed when approaching pedestrian crossings. Drivers should slow down and be prepared to stop, and on no account may they overtake any other vehicle that is slowing or has stopped to allow someone to cross on a pedestrian crossing.

Drivers must give way to any pedestrian crossing on their half of the crossing. When a school patrol sign is extended at a pedestrian crossing, vehicles travelling in both directions must stop and remain stopped until the crossing is clear and the sign at each side has been withdrawn.

PEDESTRIANS

Pedestrians are the most vulnerable of road users and they need to take extra care. They must keep to the footpath or, where there is no footpath, keep as close as possible to edge of the road.

Crossing the road should be done at right angles. If a pedestrian crossing is within 60 feet (18m) it must be used. Never step out suddenly on to a crossing when an approaching vehicle is so close that the driver would be unable to give way. Loitering is out. At traffic lights, wait for the "CROSS" signal, but if there are no pedestrian signals, wait to cross with the green light. At some specially signposted intersections, pedestrians may cross diagonally when all vehicles are stopped by the traffic lights. Last but not least — obey the signals of traffic officers.

SAFETY HELMETS AND SEAT BELTS

Laws relating to safety helmets and seat belts are relatively recent, but they are becoming increasingly far-reaching. In December 1973 it was made compulsory for all motorcyclists, power cyclists and pillion riders, except those with special exemptions, to wear approved safety helmets at all speeds.

Seat belts for the driver and front-seat passenger must be fitted to all cars, vans and light trucks first registered after January 1 1965. At the beginning of 1975 the requirement extends to all such vehicles first registered after January 1 1955.

For all cars, vans and light trucks registered after July 1 1972 seat belts must be combination (lap and diagonal) ones or full harnesses for the driver and front-seat passenger.

Particular cases may qualify for exemption from these requirements. Otherwise if your vehicle is legally required to be fitted within seat belts, you and your front-seat passenger must wear them.

Children under 15 are exempt and so are people with certificates from doctors giving medical reasons for not wearing them.

GENERAL COMMENTS

Vehicles that are dangerously loaded, annoyingly noisy or which emit excessive amounts of smoke or vapour may not be driven on public roads.

Loaded firearms must not be carried in vehicles.

Approaching traffic may be warned of a breakdown ahead by a red reflective triangle placed on the road or by the motorist using his flashing indicators simultaneously.

If broken glass, sharp objects or slippery substances fall from your vehicle, you, as the driver, are responsible for removing them.

If the driver is injured, whoever removes the vehicles is responsible for clearing the roadway.

There are a number of specific cases where you must stop your vehicle:

- When signalled by a police or traffic officer.
- When signalled by a siren (in some instances you need only make room to pass)
- For two alternately flashing red lights (railway crossings, fire stations, airports). You must remain stopped until the lights cease flashing.

Horns are to be used only as reasonable traffic-warning devices. Except in an emergency, horns must not be sounded in 50 km/h areas between 11 p.m. and 7 a.m.

At railway crossing, slow down to 30 km/h before reaching the crossing. Before you start to cross ensure that the line is clear in both directions. Do not try to cross if warning bells are ringing or warning lights are flashing.

The following illustrations are a selection of the numerous road signs, the motorist will see while travelling the New Zealand roads.

The sign showing maximum speed limits had not been gazetted at the time this book went to press. However, it is the familiar red circle enclosing a white centre on which the speed is shown in black with km/h beneath.

Warning Signs

Side road ahead crosses the railway line.

There is a pedestrian crossing ahead.

Ice may be encountered on the road in adverse winter conditions.

Road bends to the right, (or left if symbol is reversed).

Traffic roundabout ahead.

(ACCIDENT)

There is a stop sign ahead.

Road narrows from three lanes to two.

Resealing of the road is in progress, and it is sensible to travel slowly.

A temporary speed restriction, usually erected when road works are in progress.

Road Markings

The diamonds on the road warn of a pedestrian crossing ahead.

The yellow dashes mean no parking.

Informatory Signs

THERMAL AREA

RIMUTAKA SUMMIT 555 m

REST AREA 400 m

AHEADVILLE ↑
RIGHTOWN →
HAMILTON →
← LEFTHAM

NAIKE 7 km

Road Markings

The solid yellow line on your side of the broken centre line, or a solid double line means no passing.

The yellow broken lines are an advanced warning of a solid yellow line.

The motorist can travel at the maximum permitted speed.

Regulatory signs

NO TURNS

No turns are allowed off the road on which the car is travelling.

FREE TURN

At light controlled intersections, it is possible in some cases for left turning traffic to proceed and not wait for a light change.

NO U TURN

Forbidden to make a turn and drive in the opposite direction.

LSZ

The letters mean "Limited Speed Zone". It means that the motorist uses his judgement according to the density of traffic, or stock, or for any other reason, in driving at an appropriate speed.

HEAVY VEHICLES **MAX LENGTH** INCLUDING TRAILER 9ᶜ

Roads, because of some physical restriction, will not accommodate heavy vehicles over the specified length.

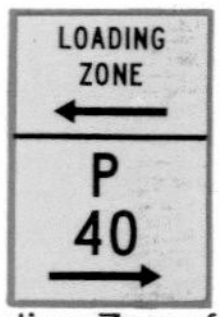

LOADING ZONE P 40

Loading Zone for service vehicles from the sign, in the direction of the arrow, but 40 minute parking from the sign in the opposite direction.

BUS STOP KEEP CLEAR 8AM TO 9AM 4PM TO 5PM P 60

No stopping for all vehicles between the hours specified, but 60 minute parking in both directions from the sign, at all other times.

NO STOPPING 4:30PM TO 5:30PM P 20

No stopping for all vehicles between the hours specified, but 20 minute parking in both directions from the sign, at all other times.

NO STOPPING DAY OR NIGHT

No stopping at all times from the sign in the direction of the arrow. This applies until another sign is met with the arrow pointing in the opposite direction.

DISCOVERING NEW ZEALAND

New Zealand was the last significant land mass discovered by Europeans in the great wave of exploration which began with Christopher Columbus at the end of the 15th century. It was not until the middle of the 17th century that a European ship sailed along the New Zealand coast; but it was no virgin uninhabited land as Abel Tasman found out when he anchored off the northern tip of the South Island. As some of his men rowed from one ship to another, they were intercepted by a canoe full of brown-skinned men and four of the crew were killed in a brief skirmish. Tasman gave the bay the name Murderers' Bay and sailed on.

Exactly when the first Polynesians arrived in New Zealand may never be known. Maori lore gives this honour to Kupe, a legendary navigator, who sailed around the coast of New Zealand and settled for a period on the Hokianga Harbour about the middle of the 10th century. Recent archaeological evidence suggests that the first settlers may in fact have arrived as early as 700 A.D., about the time that the Roman province of Britain was being over-run by invading Angles, Saxons and Jutes, and long before Scandinavian sailors made the journey across the Atlantic to North America. Yet the achievements of the Polynesian sailors who made the voyage to New Zealand, whether accidental or planned, were equal to those of the Norsemen and later explorers such as Columbus, Magellan and Cook.

The first arrivals found a land very different from the tropical islands they had left thousands of kilometres behind. In size alone the new land outstripped anything they had experienced before, and, far to the south, it was much colder than their homeland. As they soon discovered, taro and gourds would not grow in most areas of the new land, and it seems that the other staple root crop, the kumara or sweet potato, did not survive the voyage. But the loss of these staples was not too much of a handicap for the land was rich with wildlife including the giant flightless moa; and the dense, almost sub-tropical forest which covering most of the land offered sufficient sustenance to the diligent. The seas, rivers and lakes also yielded abundantly to expert fishermen. By the 12th century the moa-hunters, as these first settlers are now called, had mutliplied and spread throughout the land; remnants of their camps have been found as far inland as the shores of Lake Taupo and in central Otago.

Some time in the 13th or 14th centuries a different Polynesian culture was established in New Zealand. Whether this was imported by new arrivals, as documented in the legends of the Great Fleet, or whether it arose naturally in response to population pressures and a restriction in food supplies with the extinction of most of the moas, is not known at present. What is certain is that the new variant of Polynesian culture rapidly became dominant, although pockets of moa-hunter society survived until the 17th century. Classical Maori society was distinguished by its warrior code, the development of fortified villages or pas, and the importance of the cultivation of the kumara. Because the kumara needed warmth and sunshine, Maori settlement was concentrated in the northern half of the North Island.

Over the four or five centuries that elapsed before the coming of the European, Maori society evolved into one of the most advanced stone-age cultures in the world. The engineering skill that went into their pas won the admiration of the first European soldiers to inspect them; fierce warriors, they combined family loyalty and a strong chivalry with a fine sense of the tactical value of deceit. Partly because of the limitations of their technology, they lived in harmony with the land, respecting it as the property of the gods who allowed them to harvest its bounty, and incorporating the land into a theology more complex than the first Europeans believed or cared to understand. Today the highest levels of the Maori religious system are lost, gone with the tohungas who took their knowledge with them to the grave.

The arrival of a small sailing ship under the command of Captain James Cook in October 1769 spelt the end of Maori isolation. Over the ensuing six months Cook charted the coast with great accuracy, making frequent contact with the Maori and forming a favourable impression of the suitability of the land for settlement. Within 30 years of Cook's first voyage, other Europeans were sailing the coast of New Zealand with less idealistic motives. They were sealers who, for a few brief years around the turn of the 19th century, found a short-lived but rewarding source of wealth in slaughtering the seal colonies of the southern shores. Equally gruesome was the trade in shrunken heads that developed. The sealers were followed by others eager to exploit the new land's resources. Whalers began to call at the Bay of Islands for supplies and relaxation from months at sea; by the 1820s Kororareka in the Bay of Islands was known as the "hell-hole" of the Pacific — and with good reason. Others came to the Bay of Islands for kauri and other trees well-suited for ship-building, and for flax.

At first these new arrivals stayed only briefly, but by the 1820's traders and shore whalers had begun to settle permanently, many of them living virtually as members of tribes. The Maoris welcomed them for they were a source of the new goods they coveted, especially the musket. First to recognise the superiority the new weapon gave was the Ngapuhi chief Hongi Hika who devastated large areas of the middle of the North Island. Gradually other tribes acquired their own guns and the balance of power was restored; by the 1830s most chiefs realised that the new weapon was destroying their own people and the inter-tribal wars came to an end. But tens of thousands out of a population of approximately 250,000 had been killed and perhaps an equal number had succumbed to the diseases brought by the Europeans.

As the number of Europeans in New Zealand grew, the British Government reluctantly concluded that it would have to take the country as a colony, and in February 1840 Captain William Hobson signed a treaty with a number of Maori chiefs at Waitangi ceding sovereignty to the Crown. Over the following few months further signatures were collected throughout New Zealand, and the claim was further reinforced by proclamation. The treaty was a humanitarian document and the British intention was certainly to protect the Maori from the rapacity and land hunger of the European, but it is doubtful if the Maori chiefs understood what they were agreeing to, and the majority of colonists regarded it as little more than a worthless piece of paper.

With the growing number of colonists, Maori disquiet grew too and armed clashes between Maori and European occurred. The first incident occurred in 1843 on the Wairau Plains in Marlborough where the redoubtable Te Rauparaha confronted settlers from Nelson. In 1845 Hone Heke led a brief but prophetic campaign against the British in the north of the North Island, while in the southern part of the island, settlements are Wanganui and Wellington were also the scene of fighting.

By 1860 pressure on Maori land was becoming intense. Settlements at Auckland and New Plymouth were hemmed in by Maori reluctance to sell their land. Resentment grew especially in Auckland where the colonists eyed the rich lands of the Waikato flourishing under the European-style cultivation adopted by the Maori, but the first conflict came in Taranaki and later spread to the Waikato. By the end of the 1860s the Pakeha had triumphed and the way was clear for European expansion. The guerrilla campaigns of Te Kooti from the Urewera in the late 1860s and early 1870s, and Te Whiti's campaign of passive resistance in Taranaki, were no more than the last smouldering embers of the conflict.

While the North Island was embroiled in warfare, the South Island was forging ahead. Settlers at Nelson in 1842 found their expansion hampered by lack of suitable land, but the colonists sent out to Otago in 1848 and Canterbury in 1850 did not face such handicaps. The Maori population was small and quite willing to sell its land, and large areas were already clear of forest. As a result settlement spread rapidly, and large sheep runs covered Otago and Canterbury within a decade of their founding. In the 1860s Otago received an even greater stimulas to its development with the discovery of gold in its central region. The gold rushes lasted for less than a decade in Otago, but the wealth and population which resulted established the south as the commercial and industrial centre of New Zealand.

In the century that has passed since then, the North Island has become the dominant partner, both in terms of population and economics, but the effects of the founding years still lingers. The North Island seems to retain the urge to develop as fast as possible, as though it was still seeking to catch up with the South Island. The exploitation of the land has been more ruthless, and the new European pattern of development that has been imposed has as yet to find a balance with the older landscape. There are still many raw edges, most notably the erosion that afflicts many farming areas, especially on the East Coast, and the way hills and valleys are refashioned to allow the expansion of booming towns and cities. In contrast, the South Island appears to have grown more in harmony with the land. Even in Otago, where miners, and later dredges, ravished the earth in search of gold, the scars they left behind appear to fit into the harsher environment in which they worked. And on the West Coast most of the gold-mining areas have been reclaimed by the dense rain forest. There is a greater sense of order and knowledge that the pioneering era is firmly in the past.

It is a past which is not far distant compared to other parts of the world. New Zealand is a young country; probably less than half its population has been there for longer than one or two generations, while at the turn of the century over 90 percent of its people were first or second generation New Zealanders, and loyalty to Britain and the Empire was intense. In the two world wars of this century New Zealand supported Britain without hesitation, and it did not claim the full independence granted by the Statute of Westminster of 1932, until 1947. Yet most immigrants came with the belief that they would find a better life, and the colonial atmosphere had a liberalising effect too. The most obvious signs of these aspirations were the social welfare provisions enacted in the late 1890s and during the first Labour administration of 1935-8.

As individual memories or inherited values have weakened, so too have the ties between the mother country and New Zealand. Other factors have influenced this growing independence: the withdrawal of Britain from its position as a world power to a European one, and the growing number of Maoris and Islanders in the cities. In the remaining 25 years of this century these forces will probably cause New Zealand to forge a new identity as a Pacific and semi-Polynesian nation. Not that this role will be developed deliberately; New Zealanders are a practical people, more concerned with the solution of immediate problems than with any grand social theories — and the enthusiast has always been regarded with some suspicion. New Zealand will find its new status by experimentation, with hesitations and frictions that will perhaps in future be seen as important as the Maori-European Wars and the growth of grasslands farming.

New Zealand enters the third quarter of the 20th century with the land and people caught in a moment of transition. It is a country which is proud of its reputation for racial harmony yet must adjust to an increasingly assertive and fast-growing Maori and Polynesian population; a country which resolutely proclaims itself one of the "free" nations but is reluctant to allow expressions of some of those freedoms; a country in which over three quarters of the people living in towns and cities depend on the output of its farms for prosperity.

The land faces challenges too. Reconciling a growing population and the need for continued economic expansion with the preservation of its natural attractions will be difficult. Much that is unique about New Zealand's scenery will be lost if all its beaches, lakes and mountains are festooned with the trappings of the tourist trade. And variety is the key to New Zealand's scenic charm; the claim that it is a world in miniature is a mild — and forgiveable — exaggeration. For the New Zealanders or the visitor from overseas, the country offers a wealth of contrasts; the sub-tropical warmth and sunshine of the far north; the hundreds of kilometres of beaches; Rotorua's thermal areas; the lakes and wilderness with their superb fishing and hunting; and the grandeur of the Southern Alps.

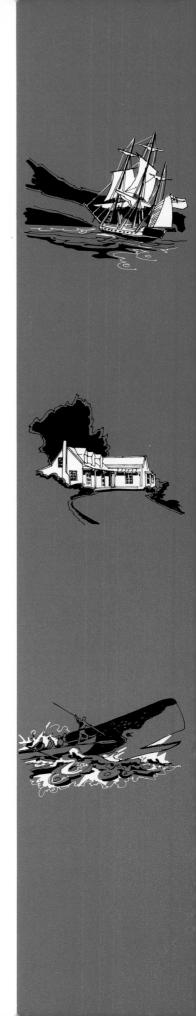

16

Native trees

POHUTUKAWA (Christmas Tree). A handsome tree confined to the Auckland Province, preferring precarious holds on cliffs overlooking the sea or lakes. At Christmas time is covered with a profusion of scarlet, nectar laden flowers.

MANUKA (Tea Tree). The most abundant of native shrubs. In forest regeneration it is the first to grow and provides nursery cover for native trees which flourish in their shelter. The leaves were used by pioneers as a substitute for tea.

TOTARA. A very light, sound and durable timber peculiar to New Zealand. A tree favoured by the Maoris for the making of canoes and carvings.

RIMU (Red Pine). Found throughout New Zealand. Used extensively for house and furniture building. Has a very attractive and distinctive juvenile form.

KAURI. One of the most magnificent timber trees known. Some specimens estimated to be 1500-2000 years old. Grows only in the Northern quarter of the North Island.

MOUNTAIN BEECH. One of the four beech trees which clothe the mountain slopes in sub-alpine areas.

MIRO. A handsome but fairly uncommon tree. It provides a bright red, plum-size berry of which the native pigeon is very fond.

KOWHAI. A widely distributed tree characterised by beautiful branches of large golden blossoms, which appear in early spring. Very popular with nectar sipping birds.

KAHIKATEA (White Pine). A swamp loving tree it is one of the tallest of New Zealand trees, specimens having reached 60m. As the wood is odourless, it has been used for packing butter, but supplies of the timber are now very limited.

SOUTHERN RATA. A relative of the Northern based Pohutukawa, is equally famed for its profusion of scarlet flowers. Massed specimens may be seen in Otira gorge blossoming in December and January.

Flowers

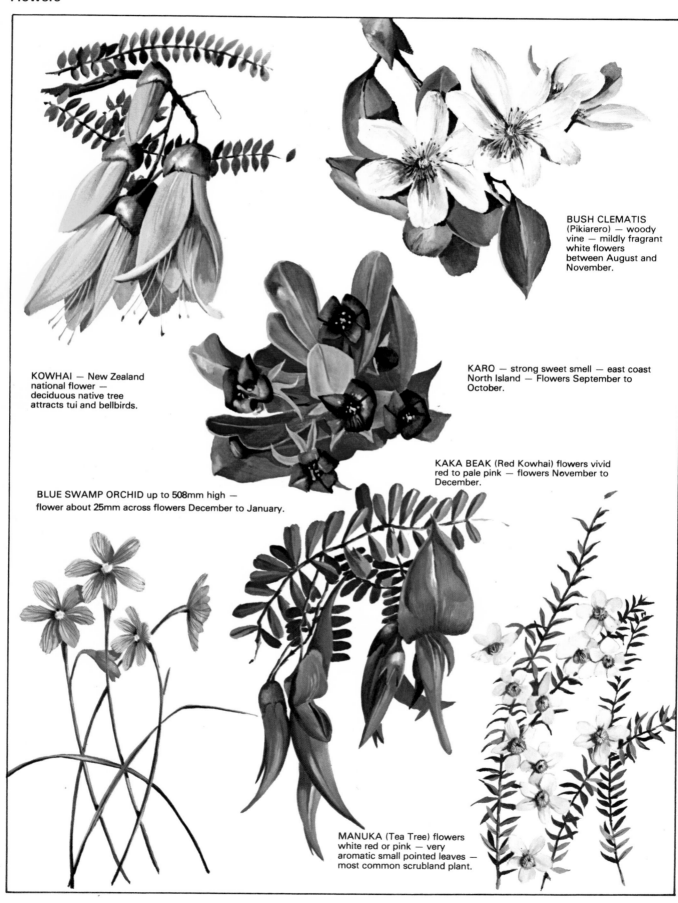

KOWHAI — New Zealand national flower — deciduous native tree attracts tui and bellbirds.

BUSH CLEMATIS (Pikiarero) — woody vine — mildly fragrant white flowers between August and November.

KARO — strong sweet smell — east coast North Island — Flowers September to October.

KAKA BEAK (Red Kowhai) flowers vivid red to pale pink — flowers November to December.

BLUE SWAMP ORCHID up to 508mm high — flower about 25mm across flowers December to January.

MANUKA (Tea Tree) flowers white red or pink — very aromatic small pointed leaves — most common scrubland plant.

NEW ZEALAND LILAC — hairy branches long serated leaves — grows in dry rocky places in north of South Island — flowers between October - December.

YELLOW ROCK DAISY — rock alpine flower — soft leaves — flowers January to March.

PINK HEBE — straggling shrub — flowers in October to March — grows on dry rocky mountain places.

BLACK BARKED MOUNT HEBE — low lying shrub growing on rocky ledges from Nelson — Canterbury up to 1371m flowers November to February.

RIMU ROA — slender stemmed plant found on grasslands — flowers September to April.

GRASSLAND DAISY — creeping herb — each stalk carries flower found alpine herbfields and grasslands — flowers November to January.

Bush Birds

A selection of the more common birds of New Zealand, the majority of which may be seen by the general traveller.

BELLBIRD
Found in forest and second growth throughout New Zealand. Eats insects, fruits and nectar. Song, a sequence of pure bell-like liquid notes. 200mm long.

PIGEON
Throughout New Zealand in forests and occasionally in exotic trees. Food, mainly berries and young leaves. A noisy flier. 500mm long.

WHITE FACED HERON
A comparative newcomer to New Zealand. Widespread in coastal districts of both islands, but scarce inland. Nests usually found high in trees.

RED CROWNED PARAKEET
Predominately a green parakeet 200-300 mm long, it's plumage affording good camouflage in green foliage. Now somewhat scarce on the mainland except in deep forest, it is plentiful on offshore islands.

SILVEREYE OR WAXEYE
Throughout New Zealand but plentiful in settled areas with tree cover. Diet consists of insects and fruit but takes nectar with enthusiasm.

BLUE REEF HERON
Distributed along the rocky coasts of main islands, off-shore and on mudflats.

KAKA
A native parrot found in trees throughout New Zealand. An insect and fruit eater. A large bird 400-480 mm long, has dramatic red colouration underwing, visible in flight. Nests in holes of trees high off the ground.

KIWI
The wellknown flightless bird now rarely seen. Lives in holes and hollow trunks in dense vegetation and only emerges at night to feed. Cry is a penetrating shrill whistle. 450-550mm long. Lays a disproportionately large egg.

FANTAIL

A very friendly little bird easily identified by its fan shaped tail. Very common in all types of forest, in public and private gardens and exotic plantations.

HARRIER

A magnificent flier seen soaring up in draughts without wing movement over open country. Will be noticed scavenging on roadsides. Nests usually in swamps or fern.

MOREPORK

Best known of the four native owls. A silent flier can occasionally be seen at dusk feeding on insects, small rodents or beetles. The call which sounds like its name, is often heard in the forest at night.

KEA

A south island bird related to the Kaka most often seen screaming around mountain tops and scree slopes. Very inquisitive and sociable. Eats mostly fruits, buds and leaves obtained from forest below the snowline. Nest among rocks or in a hole in the ground.

TUI

A handsome bird with a voice resembling the bellbird. Can often be seen in the forest flying noisily or moving vigorously among the trees. Eats insects , keen on nectar.

SOUTH ISLAND ROBIN

An inquisitive, fearless little bird, long-legged and slate-coloured with pale underparts. Feeds on insects and grubs on the forest floor.

KINGFISHER

Unmistakable with a spectacular colour scheme, this bird is often seen perched on fence posts and telegraph poles. Inhabits settled areas, lake shores, estuaries, rivers

PUKEKO

A handsome swamp rail. Prefers walking to flying, but will fly if pressed. Nests in and inhabits swamps. Emits an ear piercing scream, often heard at night.

Water Birds

YELLOW EYED PENGUIN
A sedentary species recorded in the South Island and southward from there. Nest in scrub shelter and low forest. 760 mm high.

RED BILLED GULL
The common scavenging gull seen around the coasts and on some inland lakes. Densely packed nesting colonies may be seen on sandbanks, reefs and cliffs.

BLACK BACKED GULL
A large scavenging gull found throughout the country. Frequently seen in river beds inland lakes and alpine tarns. Breeding grounds include outlying islands, reefs.

WHITE FRONTED TERN
A graceful sea-swallow commonly seen throughout New Zealand. Often seen in flocks diving into shoals of tiny fish which are in turn being chased by schools of Kahawai. Breeding grounds are seen in sand dunes, off-shore inlets and estuarine shingle banks.

PIED SHAG
More plentiful in the North of the North Island and also seen in the North of South Island. Nests in trees especially pohutukawa, overhanging cliffs.

CASPIAN TERN
With a wingspan of 1300 mm it is the largest tern. Fairly plentiful around New Zealand and often seen inland and in rivers and lakes. Nests on shingle banks, sand dunes .

PIED STILT
An unmistakable black and white wader with long slender pink legs. Frequently heard calling during migrating night flights with sounds similar to a dog yapping. May be seen inland in swampy paddocks lagoons, and lakes.

SPOTTED SHAG
A handsome 990 mm cormorant. A sedentary bird it nests on cliff ledges handy to deep water. Not common in the North Island but abundant in the South Island.

GANNET

A majestic coastal bird noted for dramatic dives for food. Usually breeds on offshore islands, but an exception is a huge land based colony at Cape Kidnappers.

DABCHICK

Belongs to the grebe family, is plentiful in the North Island but scarce in the South. Is seen inland on lakes and inland rivers.

VARIABLE OYSTERCATCHER

A lonely and rather morose looking coastal inhabitant. Can be seen stalking about sometimes in pairs on beaches estuaries, and rock pools. Found in both islands.

BANDED DOTTEREL

An unmistakable attractive little plover seen defending their beach nests after Christmas, through most of New Zealand. Some migrate to Australia for the winter.

GREY DUCK

Brown plumage rather than grey. Universal distribution throughout New Zealand. Seen feeding wherever there is water but tends to nest away from the water.

NZ DOTTEREL

Breeds in two widely separated localities — the North and the far south. A sedentary species it frequents ocean beaches, sandy flats and estuaries.

WHITE SWAN

A handsome introduced swan breeding in considerable numbers on Lake Ellesmere. Builds a massive nest at the edge of swamps.

GODWIT

This tidal flat bird breeds in N.E. Siberia and migrates annually to New Zealand arriving in September It can be seen assembling in huge flocks in late summer on the large northern harbours for the return flight. KNOT: A small shorebird seen often in large flocks. As with the Godwit, it breeds in Siberia and is frequently seen with them. Distributed throughout New Zealand though most common in the North.

Fish

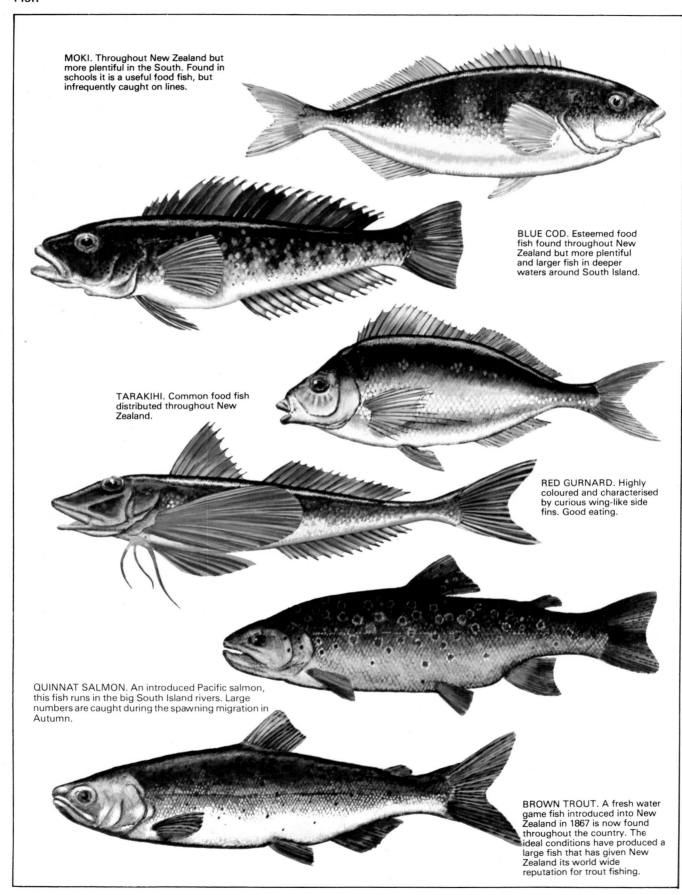

MOKI. Throughout New Zealand but more plentiful in the South. Found in schools it is a useful food fish, but infrequently caught on lines.

BLUE COD. Esteemed food fish found throughout New Zealand but more plentiful and larger fish in deeper waters around South Island.

TARAKIHI. Common food fish distributed throughout New Zealand.

RED GURNARD. Highly coloured and characterised by curious wing-like side fins. Good eating.

QUINNAT SALMON. An introduced Pacific salmon, this fish runs in the big South Island rivers. Large numbers are caught during the spawning migration in Autumn.

BROWN TROUT. A fresh water game fish introduced into New Zealand in 1867 is now found throughout the country. The ideal conditions have produced a large fish that has given New Zealand its world wide reputation for trout fishing.

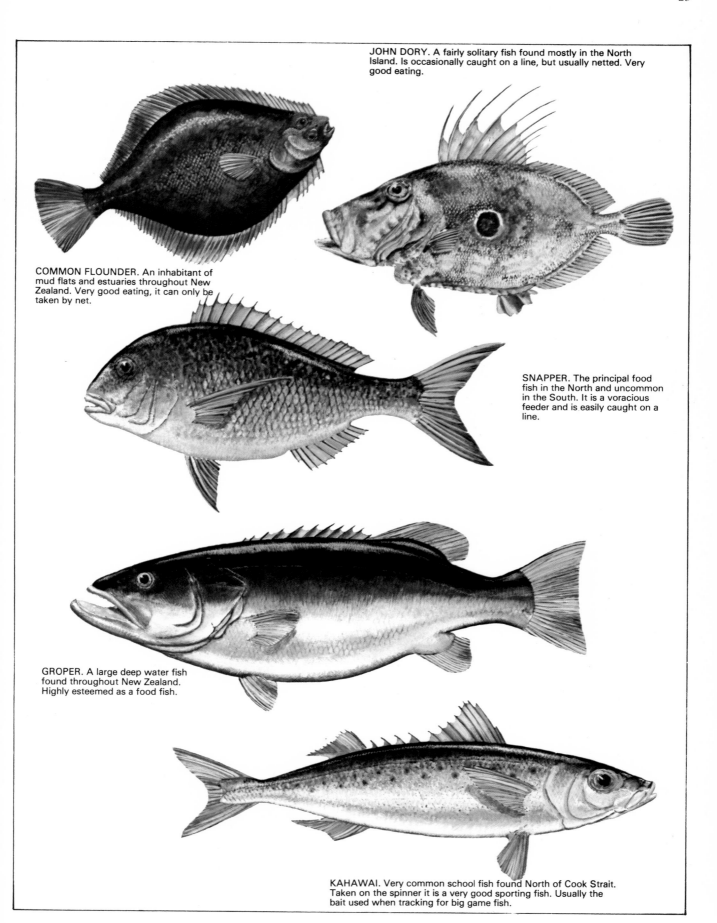

JOHN DORY. A fairly solitary fish found mostly in the North Island. Is occasionally caught on a line, but usually netted. Very good eating.

COMMON FLOUNDER. An inhabitant of mud flats and estuaries throughout New Zealand. Very good eating, it can only be taken by net.

SNAPPER. The principal food fish in the North and uncommon in the South. It is a voracious feeder and is easily caught on a line.

GROPER. A large deep water fish found throughout New Zealand. Highly esteemed as a food fish.

KAHAWAI. Very common school fish found North of Cook Strait. Taken on the spinner it is a very good sporting fish. Usually the bait used when tracking for big game fish.

Shells

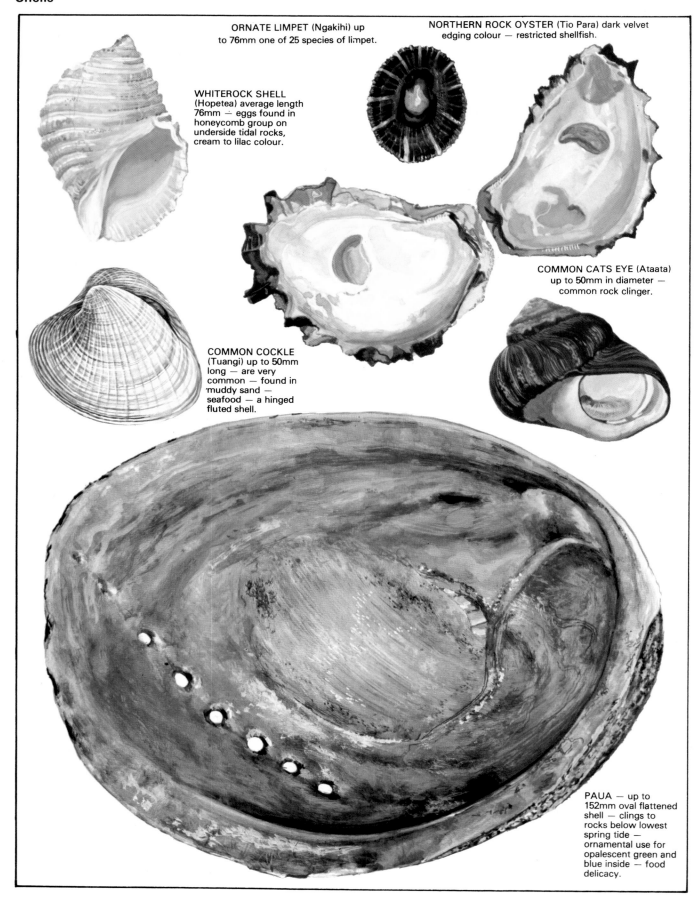

ORNATE LIMPET (Ngakihi) up to 76mm one of 25 species of limpet.

NORTHERN ROCK OYSTER (Tio Para) dark velvet edging colour — restricted shellfish.

WHITEROCK SHELL (Hopetea) average length 76mm — eggs found in honeycomb group on underside tidal rocks, cream to lilac colour.

COMMON CATS EYE (Ataata) up to 50mm in diameter — common rock clinger.

COMMON COCKLE (Tuangi) up to 50mm long — are very common — found in muddy sand — seafood — a hinged fluted shell.

PAUA — up to 152mm oval flattened shell — clings to rocks below lowest spring tide — ornamental use for opalescent green and blue inside — food delicacy.

CUNNINGHAMS TOP SHELL — up to 50mm long — sandy shores mainly West Coast — fawn colour with spiral ribbing.

TIGER SHELLS (Maurea) up to 76mm diameter — found on open rocky coasts — not common.

NORTHERN SIPHON WHELK (Kawari) 50-63mm in length carnivorous — often found clinging to cockle — white speckled with brown-yellow inside opening.

HELMET SHELL up to 63mm long — open sandy beaches — small knobs around shoulder.

ARABIC VOLUTE (Pupu Rore) up to 152mm long — found below low tide — North Island sandy beaches.

SILVER PAUA (Hihiwa) smaller than paua — silvery interior up to 101mm long.

COARSE DOSINA (Tuangi Haruru) up to 88mm thick, disc shaped, concentric ridges — buries in sand below low water mark.

Introduced Animals

RABBIT — introduced before 1860 for skins and meat. They now ruin vegetation and soil and affect stock holding capacity on farmland — Rabbit Control Boards aim to exterminate the pests.

OPOSSUM — an Australian marsupial introduced for skin trade. Found throughout New Zealand they damage forests, orchards, and gardens. A ''noxious animal'', poison is widely used to reduce their numbers.

RAINBOW TROUT — originally introduced from California differ from the brown trout in that they have a reddish band along side of the body. Spawning takes place in rivers but they return to the lakes to feed. Well conditions rainbow average about 4 lbs.

HARE — introduced for sport these animals live between sea level and 6,000 ft. generally in grassland areas. Exports of skins and meat to Europe mainly from the South island. Hares damage young trees, grass, orchards and gardens. Poison and shooting are main extermination methods.

RED DEER — most common species — inhabits mountain regions between Bay of Plenty and Stewart Island — forest grassland and scrub — cause damage to vegetation.

STOAT found mainly in open country and forests. Introduced to control rabbits there is no control programme in operation.

HEDGEHOG controls garden pests, snails, slugs, etc. It is thought that hedgehogs eat eggs of ground nesting birds.

WEASEL belonging to same family as stoat, the weasel is much smaller and has no black tip on its tail.

PHEASANT. Two species were introduced, a chinese ring necked pheasant and an English variety. Numbers controlled by sportsmen and poisoned bait.

AUSTRALIAN BLACK SWANS found mainly in shallow lakes and lagoons where there is a plentiful supply of water and swamp weeds.

MYNA. An Indian bird found mainly in north of New Zealand.

SUFFOLK cross of Southdown rams and Norfolk horned ewes — meat and wool.

BORDER LEICESTER cross English Leicester and Cheviot ewes — meat.

MERINO originally a Spanish breed — a great wool producer.

CORRIEDALE — Lincoln and Leicester rams with Merino ewes — meat and wool

ROMNEY MARSH developed in Kent — thrives in wet areas — meat and wool.

CHEVIOT a wool producer from mainly low country farms.

SOUTHDOWN an early maturing mutton—producing breed.

Dairy and beef cattle

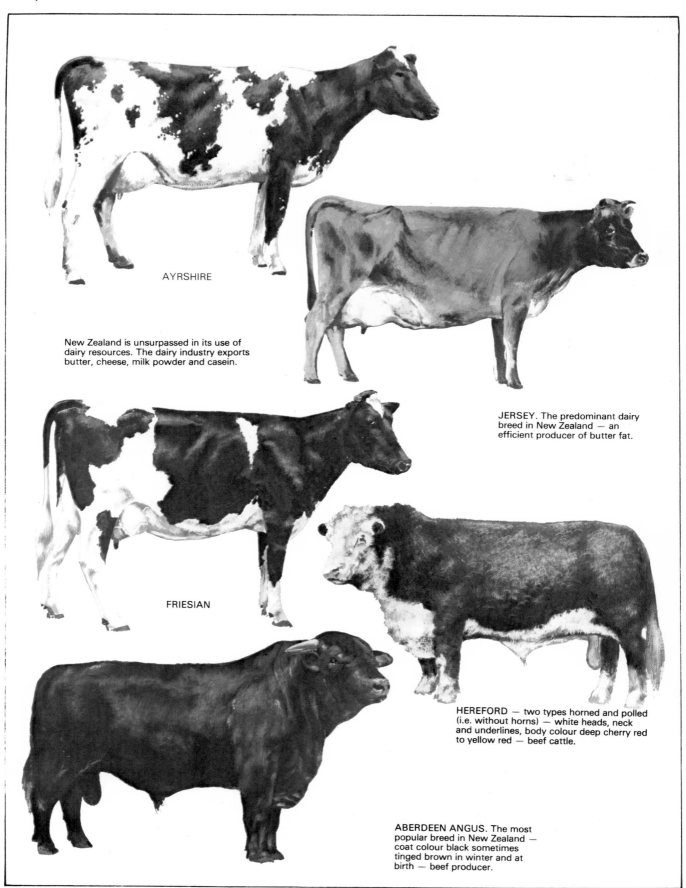

AYRSHIRE

New Zealand is unsurpassed in its use of dairy resources. The dairy industry exports butter, cheese, milk powder and casein.

JERSEY. The predominant dairy breed in New Zealand — an efficient producer of butter fat.

FRIESIAN

HEREFORD — two types horned and polled (i.e. without horns) — white heads, neck and underlines, body colour deep cherry red to yellow red — beef cattle.

ABERDEEN ANGUS. The most popular breed in New Zealand — coat colour black sometimes tinged brown in winter and at birth — beef producer.

ECOLOGY & CONSERVATION

The environment and the car

Man is as much a part of nature as the sun, the air, water, soil and all other living things.

But we have developed a way of life quite unlike that of the natural world. As a result we have global problems of pollution. Resources are diminishing or depleted. Our urban living seems to have more problems than advantages — congestion, crowding, a sense of isolation and meaninglessness, leading to crime and neurosis.

But the natural world and the manmade should not be in opposition. Both determine our environment and unless we integrate our way of life with that of nature's, we will fail as human beings.

We must build a way of life which does not destroy irreplaceable life systems, which does not involve waste of people and resources, which does not gradually destroy bush, forests and land, and we must build an urban environment which meets the non-material needs of people as well as the material needs.

Already we have begun to put out new understanding into action. The system of water rights and the Clean Air Act are restoring the health of our water and air. The Town and Country Planning Act has been revised.

We have begun to find out how to conserve and recycle resources; we have begun to recognise that the urban sprawl of cities wastes land, isolates people and affects the quality of our lives.

But in trying to find ways to solve our problems, we are faced with a dilemma — our material wants clash frequently with our non-material needs.

The private car is the symbol of this dilemma in our modern world. And it is also the cause of many of the problems now threatening not only the quality of life, but survival itself.

When cars were first developed they were welcomed with open arms. They did not seem to have safety or control problems. They did not need professional drivers. They were an ideal consumer product giving the industrial world the chance it wanted to expand.

And we began to build our life around them. It was no longer necessary to live near work — so we moved to suburbs and built motorways to reduce the time it took to travel the extra distance.

With the car came the chance to experience our own strength, skill and speed. It gave a sense of power — over sophisticated machinery, and over people. From it came communication and contact with others. Driving gave freedom and privacy. And along with it came fringe benefits —

— a sense of being prosperous and living comfortably;
— a sense of life as a game of sportsmanship;
— status and prestige;
— virility — if we are to believe the ads!

It is hardly surprising that we found adverse side effects, deaths, spoiling of the rural and urban environment, tolerable for a while. It could be expected that we would assume every man and woman would want to own a car in the future and want higher wages to make it possible.

But now the harm to our environment is quite clear; car exhausts pollute, cars kill and maim, deplete our resources, make exhorbitant demands on energy; assembling them makes monotonous and dull work for many people. The motorways and car parks pave over land and make it useless for anything else. With more and more people owning cars the car is no longer just a friend. It has instead the potential to be a deadly enemy. The prospect of New Zealand ever reaching a saturation level of car ownership is chilling.

Because with the awareness of the global problems of pollution, of resource depletion, of the fact that our urban environments are becoming unsuited to people, of despoliation of the countryside, has come a recognition that with the benefits of cars come serious drawbacks —

— dangers to the life and health of man, animal and plant;
— disruption of ecological equilibrium;
— aggressiveness, noise, intolerance;
— destruction of the beauty of the land from roadside hoardings, from flag-ridden petrol stations;
— gross wastage of energy and other resources;
— a loss of freedom for pedestrians;
— frustrating delays on roadways and in parking lots.

We have moved from the stage where cars had the advantage of speed, through the stage where we designed our towns and cities to suit car ownership and adjusted them to fit in pedestrians; to a stage where we now want to preserve the desirable features of life; to make sure the environment is at the very least tolerable and hopefully pleasurable; we want to use resources, particularly energy, as wisely as possible.

The private car is no longer the only form of transport we want. Now for much of our travelling, effective public transport would be better — express buses, trams, automated drive and fare collection systems, minibus services, pick-up routes.

But no-one wants to do away with the private car altogether— it does still give freedom of movement and privacy. It does still let us see other places in cities and the country about which we are curious, and when we know our country better it is easier to appreciate it. Driving itself is pleasurable for many people.

But clearly, if New Zealand is to have both the car and a healthy, vital, natural and manmade environment we have to learn quickly and well to develop a different kind of car and to use existing cars with some degree of sanity and plain commonsense.

And some of the things every person can do are:

— press for research into and development of 'clean' engines, new fuels.
— insist on public transport where public transport should be; on small vehicles and bicycles for town and city use;
— work to develop town planning which gives the chance to live, work and play without travelling long distances;
— press for flexible working hours to give more economic use of public transport systems and to reduce 8-9/5-6 congestion and frustration;
— avoid using your own car if you don't really need to or don't have a special reason for doing so;
— recognise why you want a car — it really doesn't give you prestige or enhance virility. Buy a horse-power rating you need — no more;
— use your car like the sophisticated machinery it is — have a tune-up regularly.
— keep it quiet as possible, and help reduce noise pollution. Get damaged mufflers and tail pipes fixed
— practice fuel economy. Turn off the engine when you stop (exhausts are pollutant anyway). Don't accelerate rapidly — imagine an egg between your foot and the accelerator.
— recognise that traffic rules are trying to make the roads easier and safer for everyone. Keep to moderate speeds and respond to the rules.
— most of all, when you do use your car get maximum satisfaction from your travelling. Use good road maps, strip maps, and New Zealand travel books. Don't miss the places of beauty, of historic interest, of sheer pleasure.

NATIONAL AN

Ten national parks have been constituted in New Zealand, covering 2,023,430ha, one-th
of the country's land area. They are administered under legislation designed "for the pu
preserving in perpetuity as National Parks, for the benefit and enjoyment of the public,
New Zealand that contain scenery of such distinctive quality or natural features so bea
unique that their preservation is in the national interest".

New Zealand's national park system had its beginning only 15 years after the worl
national park — Yellowstone, in the United States — was established. In 1887 Te Heuheu
and other Maori chiefs presented to the Crown the land within a radius of 2km of three
peaks in the centre of the North Island — Ruapehu, Tongariro, and Ngauruhoe. These pea
tapu (sacred) to the Maori people, and the gift was made on the condition that the land w
preserved as a national park. In 1894 the Government passed legislation constituting th
peaks and surrounding land as Tongariro National Park.

Six years later Egmont National Park was constituted, and in 1904 Fiordland was procl
national reserve, administered under the Scenery Preservation Act and the Tourist and
Resorts Control Act. Arthur's Pass and Abel Tasman Parks were established by the
Reserves, Domains, and National Parks Act 1928, so that by 1952 there were five nationa
— Tongariro, Egmont, Abel Tasman, and Arthur's Pass — administered by park boar
Fiordland, a public reserve administered by the Department of Lands and Survey.

Increased post-war interest led to a review of park administration to achieve unifo

Coromandel Forest Park (66,459ha) located 130km S.E. from Auckland, 112 km N.E. from Hamilton. It has a history of kauri logging and mining, also dams and other relics of bygone years. It also contains the Manaia Kauri Sanctuary. Activities include picnicking, camping, tramping, hunting, (goats and pigs), and rock hounding. Permits are required for hunting and removal of gemstones. Ranger in charge, Kauaeranga Valley, via Thames.

Auckland ●

Pirongia Forest Park (12,891ha.) located 32km S.W. from Hamilton. The central feature is the 959m forest clad Mt Pirongia, which is an old volcanic cone and has a walking track to the summit. Part of the park is water supply area where entry is restricted. Activities include hunting, tramping, picnicking and camping is permitted in most areas.

Kaimai—Mamuku Forest Park (42,800ha.) located 15-km N.W. from Tauranga, 60km N.E. from Hamilton. A ridge-top park embracing most of the remaining indigenous forest of the Kaimai and Mamuku ranges. Highest point is Mt Te Aroha (950m) and is the southern limit of kauri in eastern North Island. It contains many historic range crossing routes and logging tracks, also the 1600ha. Ngatukituki kauri/beech sanctuary. Activities include tramping, picnicking, gemstone collecting and hunting for goats and pigs.

● Hamilton

Kaimanawa Forest Park (76,042ha.) located immediately south of Lake Taupo and east of Tongariro National Park. Extensive forested and open tussock tramping country to 1700m mostly untracked. Contains both sika and red deer and productive trout streams. Snow may fall at higher levels at any time of the year. Activities include hunting, tramping, scenic drives, camping and fishing. Ranger in charge 11 Laughton St, Taupo.

Urewera National Park (199,421 ha) surounds Waikaremoana and is rich in Maori history. It largest remaining area of unspoiled native forest North Island, Kiwi, kaka, and most other native bir found throughout.

Egmont National Park (33,298ha) contains one of the world's most symmetrical mountains, the extinct volcanic cone of Mount Egmont (2517m), and includes land within a 10km radius of the mountain. It varies from heavily forested lower slopes to the bare scoria, rock, snow, and ice of the upper levels.

● New Plymouth

Kaweka Forest Park (60,827ha.) located 65km N.E. of Napier. Extensive foothill afforestation with untracked indigenous forest and open tussock country to 1700 metres. Contains red and sika deer and has snow in winter. Activities include hunting, tramping, camping and trout fishing.

Napier ●

Tongariro National Park (66,110ha) includes three volcanoes — Mount Ruapehu (2796m), the highest in the North Island and active at times; Ngauruhoe (2290m), constantly active; and Tongariro (1968m), mildly active. Mount Ruapehu is the ski-ing playground of the North Island. These regions are surrounded by native forests, open grassland, and sub-alpine vegetation.

Ruahine Forest Park (93,433ha.) located about 20km N.E. of Palmerston North and 50km W. of Napier. Extensive and often rugged mountain system of deep valleys and high points up to 1730m. Contains deer, goats, pigs and also brown or rainbow trout. Activities include hunting, tramping, fishing, camping and photography.

Tararua Forest Park (110,652ha.) located 48km N. of Wellington, extending N.E. to the Manawatu gorge. Extensive forested tramping country to 900m with open tussock tops to 1500m. Developed roadend recreational camping and picnic sites at Kiriwhakapapa and Mt Holdsworth on the eastern side and north Manakau on the western side of the park Severe climatic changes are liable at short notice. Activities include camping, fishing, ski-ing, hunting, short distance bush walks and arduous high altitude traverses.

Rimutaka Forest Park. (14,050ha) located 32km E. of Wellington stretching from Cook Strait to the Tararua Range. On the extreme west a small portion of forest (Catchpole) provides camping and picnic facilities and is the commencement of the "5-mile track" and access to the popular Orongorongo Valley and the Forest Park beyond. Activities include tramping, hunting, bush walks and camping.

● Wellington

Haurangi Forest Park (13,811ha) located about 25km S.W. of Martinborough, terminating at Cape Palliser in the S. It contains beech and mixed forest rising to about 965m mostly untracked. Activities include hunting and tramping.

REST PARKS

and management. As a result, the National Parks Act of 1952 was enacted which repealed ion affecting existing national parks, brought them all under the new Act, and established ational Parks Authority, representative of private organisations and Government ments with interest in the parks, to exercise general control which is almost entirely ment financed. Its task is to preserve parks in their natural state and so to administer at the public, in the words of the National Parks Act, "may receive in full measure the in- n, enjoyment, recreation, and other benefits that may be derived from mountains, forests, , lakes, and rivers".

park is under the control of a separate park board which carries out development and s private enterprise to establish amenities for the park users. Suitable parts are set aside erness areas where, to ensure the preservation of the natural state, development is ed to access by foot tracks. Supervisor of National Parks assists the Authority and perma- rk rangers appointed by the boards are responsible for development, protection, and in- ation of the parks to visitors.

headquarters are being established in each park as information centres and to house al displays and exhibitions of flora and fauna. Many of the parks have alpine gardens, and ature walks have been established with named trees and shrubs.

new parks have been constituted under the 1952 Act — Mount Cook (1953), Urewera , Nelson Lakes (1956), Westland (1960), and Mount Aspiring (1964).

Abel Tasman National Park (17,782ha) contains coastal and elevated bush-clad country along the shores of Tasman Bay and some off-shore islands and reefs. It is a maritime park with broken coastline, numerous bays, coves, and beaches of golden sand.

North-West Nelson Forest Park (357,865ha) located 65km due W. of Nelson. Vast forested and open tramping country including the historical Heaphy and Wangapeka tracks. Interesting limestone caves and rock formations and includes the Hikutau Forest sanctuary. Activities include rock collecting, hunting (permits required), tramp- ing, camping and fishing.

Nelson Lakes National Park (57,114ha) is centred on the twin lakes, Rotoiti and Rotoroa, and is surrounded by mountainous country with lower slopes covered by mainly beech forests. The slopes of Mount Robert have extensive snow fields.

Arthur's Pass National Park (98,296ha) is an alpine and forested region taking in part of the Southern Alps con- taining the headwaters of the Waimakariri and Otira Rivers. The transalpine road connecting Canterbury and Westland crosses the Pass and a railway runs beneath through the 8km Otira Tunnel.

Lake Sumner Forest Park (73,578ha) located 124km N.W. of Christchurch. Central feature is the 1425ha lake which adjoins the park and contains brown trout; also pic- turesque mixed beech forest especially in the Hope and Hurunui Valleys. It has a number of peaks over 1500m and recognised routes crossing the Southern Alps into Westland. Activities include tramping, camping, fishing and hunting.

Westland National Park (85,092ha) extends from sea level to heights of more than 3500m on the western side of the Main Divide. The park contains extensive native forests, lakes, rivers, and waterfalls. Features include the Fox and Franz Josef Glaciers, and Lakes Matheson, Mapourika, and Wahapo.

Craigieburn Forest Park (4451ha) located 112km west of Christchurch. Two-thirds of parks is open tussock or rock mainly used for winter sport. It has drive-in access to high altitude car parks and ski tows and is used in summer by trampers and sightseers. Used also for ex- perimental high altitude revegetation, hydrological and climate studies. Activities include ski-ing, tobogganing, and hunting.

t Aspiring National Park (158,834ha) extends wards along the Southern Alps from Haast Pass at the head of Lake Wanaka to the boundary of and National Park at the head of Lake Wakatipu. A of mountains, forests, and rivers, the dominating e is the 3036m Mount Aspiring.

Mount Cook National Park (70,000ha) lies south from Arthur's Pass and at a higher altitude. With the Main Divide its western boundary, it contains 15 peaks above 3048m among them the highest in New Zealand, Mount Cook (3764m). One-third of the total area is make up of permanent snow and glacier ice. The largest glacier is the 29km-long Tasman Glacier, which is up to 3km wide.

Fiordland National Park (1,214,000ha) is the largest park. It is an area of majestic scenery with mountains, forests, lakes, fiords, and bush-clad islands. It includes Lakes Manapouri and Te Anau, while a road through the Homer Tunnel gives access to Milford Sound. The Sutherland Falls (580m) are the highest in New Zealand. The park is the only known habitat of the flightless notornis or takahe, and the kakapo, a species of native parrot.

● **Nelson**

Christchurch ●

● **Dunedin**

● **Invercargill**

Catlins Forest Park (40,500ha) located 80km E. from Invercargill, nearest town is Owaka. Extensive untracked areas of mainly silver beech forest rising to approximately 720m. Some formed roads to boundaries and one through loop road from Owaka. Activities include tramping, hunt- ing, picnicking and some trout fishing.

AUCKLAND

From its birth Auckland seems to have been destined to develop a character and atmosphere different from any other New Zealand City. Even before William Hobson selected the Tamaki Isthmus as the site for New Zealand's capital in July 1840, eager entrepreneurs were gathering on the islands of the Hauraki Gulf in expectation of the profits to be gained from buying and selling land. This pursuit of financial and commercial success has remained the most important influence shaping the city's growth.

This of course is only one aspect. Aucklanders have an enthusiasm, openness and belief in the quality of their way of life perhaps unequalled throughout the rest of New Zealand. Beneath their hard-headed facade they are proud of their city and ready to welcome anybody willing to adopt it as their own. As a result, Auckland has proved a magnet not only for people from throughout the rest of New Zealand but also for immigrants from Britain, Europe and especially the Pacific Islands.

The mixture of racial and cultural groups has given an unique diversity and international flavour. It has also brought problems as Auckland adjusts to its role as the largest Polynesian city in the world; but from this process of accommodation a new depth and maturity will emerge.

Auckland is a sportman's paradise. With its extensive harbours and warm climate, it is a haven for yachtsmen and power-boat owners; the Anniversary Day Regatta which is held every year in late January is the largest one-day yachting regatta in the world.

Geography

The centre of Auckland lies on the Tamaki Isthmus, the narrow strip of land — in some places no more than one mile across — which links Northland with the rest of the North Island. On the east coast the Waitemata Harbour, an extension of the larger Hauraki Gulf, provides a safe deep-water port; on the opposite coast and lying a little more to the south, Manukau Harbour makes an even greater indentation.

Auckland's landscape is dotted with the remnants of approximately 60 old volcanoes, reminders of the natural forces which shaped the land over millions of years. One of the youngest is Rangitoto, the island which seems to guard the entrance to the Waitemata Harbour; scientific research has established that Rangitoto may have erupted as recently as 750 years ago.

Climate

One of the four largest cities, Auckland generally enjoys the warmest and most equable weather. Sunshine averages 2,090 hours each year with temperatures over 23°C in summer and a mean average of 8°C in the coldest month, July. Frosts are infrequent and snowfall is unknown.

Rainfall is spread fairly evenly through the year. Auckland has a high annual average of 1,243mm with June and July as the wettest months. The prevailing wind is from the south or south-west, with gales of over 64km/h on 46 days of the year.

The greatest drawback to Auckland's climate is the high humidity which can become trying in February and April. However there is no doubt that this contributes to the abundance of "sub-tropical" vegetation and to the ease with which citrus fruits, tamarillos, kiwi fruit and even bananas and pawpaws can be grown.

History

Auckland's site has been a favoured location since the first days of human occupation. To the Maoris it was known as **Tamaki-makau-rau**, "the maiden contested for by a hundred lovers". Even in pre-European times it was a densely populated area; most of the larger volcanic cones in the region bear extensive marks of ancient fortifications, and it has been estimated that the Pa on One Tree Hill once had a population of approximately 5000.

The Rev. Samuel Marsden, the noted early missionary, was the first European to explore the area when he visited briefly in December 1820. Almost 20 years later, when it became the capital of the newly-founded British colony, the area was virtually unpopulated, its Maori inhabitants decimated by the fighting which had culminated in the musket wars of the 1820s. In September 1840 Lieutenant-Governor William Hobson purchased 3,000 acres from the surviving Nga Marama and named the new town Auckland in honour of his patron the Earl of Auckland, then Governor-General of India.

Auckland's reign as capital was comparatively brief; in 1865 the government was transferred to Wellington, but this loss and the disruptions of the Maori-European Wars did little to hamper growth. By 1871 Auckland had become a city and by the turn of the century, having successfully weathered the depression of the late 1870s and 1880s, it was well on the way to the commercial and industrial dominance it has enjoyed ever since.

Population

Estimated as at April 1, 1973: 747,339

Points of Interest

City Views

Mt. Eden a large, extinct volcano, with remains of a Maori fortified Pa, offers perhaps the best vantage point for viewing the city.

One Tree Hill, Auckland's largest volcano and a splendid example of a Maori Hill Pa, offers superb views of the city. On the summit is the grave of John Logan Campbell, the "father" of Auckland, who erected the stone obelisk and gave Cornwall Park to the people of Auckland.

Museums and Art Galleries

Auckland War Memorial Museum, Auckland Domain, has one of the finest collections of Maori and Polynesian relics and carvings in the world. The display of New Zealand Birds, and the Planetarium, are also noteworthy. The Museum commemorates the two World Wars, and the War Museum on the top floor includes artillery and aircraft.

City Art Gallery on the corner of Wellesley and Kitchener Streets. Collection of European masters, 20th century sculpture and New Zealand art. Hours: Monday — Thursday, 10 a.m. to 4.30 p.m.; Friday, 10 a.m. to 8.30 p.m.; Saturday, 10 a.m. to 4.30 p.m.; Sunday 2 p.m. to 4.30 p.m.

Public Library, Lorne Street. A large modern library which includes in its collection a remarkable number of rare books and manuscripts. Hours: 9.30 a.m. to 8 p.m. Monday — Thursday, 9.30 a.m. to 9 p.m. Friday. Hours for specialised departments vary.

Museum of Transport and Technology At Western Springs off the Great North Road, close to the zoo. Exhibits of historical and technical interest in the fields of aviation, agriculture, communications, photography, road and rail transport and colonial pioneering. Hours 10 a.m. to 5.30 p.m. daily.

Buildings

City Administration Building on the corner of Greys Avenue and Cook Streets. View of Auckland from city's highest building. This is the first new municipal building in the proposed Civic Centre, being constructed around it. The Town Hall nearby (opened 1911) contains the City Council Chambers, main Town Hall and Concert Chamber.

St. Stephens Chapel erected in 1857 in a charming setting beside Judges Bay, replaces an old stone church consecrated by Bishop Selwyn.

Bishops Court in St. Stephens Ave., Parnell, including 'Selwyn Court'. Built in 1863 to the design of Frederick Thatcher, noted early architect. St. Marys Procathedral nearby is unusual in that it captures the spirit of Gothic architecture in wood. the new **Holy Trinity Cathedral** will be the centre of Anglican worship in Auckland.

University of Auckland (1882) 10,000 students frequent the beautiful new buildings. The University offers a wide range of subjects, including Schools of Medicine, Law, Engineering, Architecture, Music and a wide range of subjects, including Schools.

Gardens and Scenic Reserves

Winter Gardens offers a wide variety of tropical and sub-tropical plants in an attractive indoor setting. Hours: 10 a.m. to 12 noon, 1 p.m. to 4 p.m. Set in **Auckland Domain,** a pleasant expanse of park and sports fields.

Albert Park above Victoria Street East. A pleasant retreat from the city bustle, with some fine old trees and attractive gardens.

Zoological Park at Westmere. In colourful gardens are 300 mammals, birds and reptiles, a children's zoo, an aquarium and a zoo train. Refreshments available in the grounds. Open 9.30 a.m. to 5 p.m. daily.

Cornwall Park around One Tree Hill. City's largest park with over 130 ha of trees and parkland. A rather unique patch of farmland in city centre.

Parnell Rose Garden in Gladstone Road, Parnell. Worth visiting at any time the year especially between November and March.

Ellerslie Race Course at Ellerslie. Offers over 12 ha. of gardens as well as very popular racing facilities.

Auckland Centennial Park, Waitakere Ranges, Western Auckland. Offers magnificent native bush including Kauris, and lovely views of the city from the Scenic Drive. Refer to the AA's Waitakere Ranges map for detail of the many tramping tracks and bush walks. Cascades Park and Kauri Park (access through Swanson) are well worth a visit.

General

Orakei Marae. Off Tamaki Drive. The home of the Ngati Whatua tribe since about 1837, this is now the Marae for all the people of Auckland. A large meeting house has been built, and a cultural and recreational complex is under construction. The Marae is doing much to promote racial harmony in Auckland.

Hauraki Gulf. Ideal conditions for all types of small boating with many islands as an additional attraction. Some of the islands are part of the Hauraki Gulf Maritime Park.

The City is fringed with numerous beaches ideal for bathing or picnicking — especially those from Judges Bay to St. Heliers along Tamaki Drive — and on the North Shore, the East Coast Bays from Devonport north to Long Beach. For additional information on boating and picnicking and recreation in the Auckland district, refer to the AA's "Auckland District Map" and the Auckland City Map.

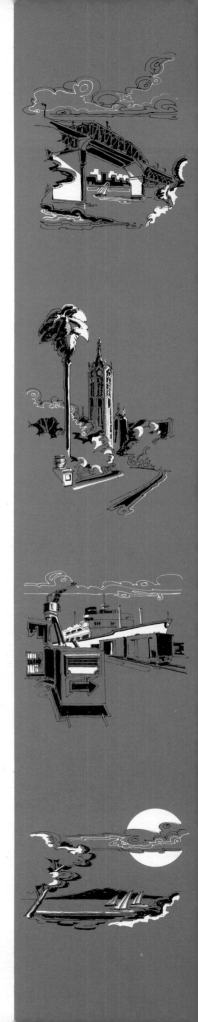

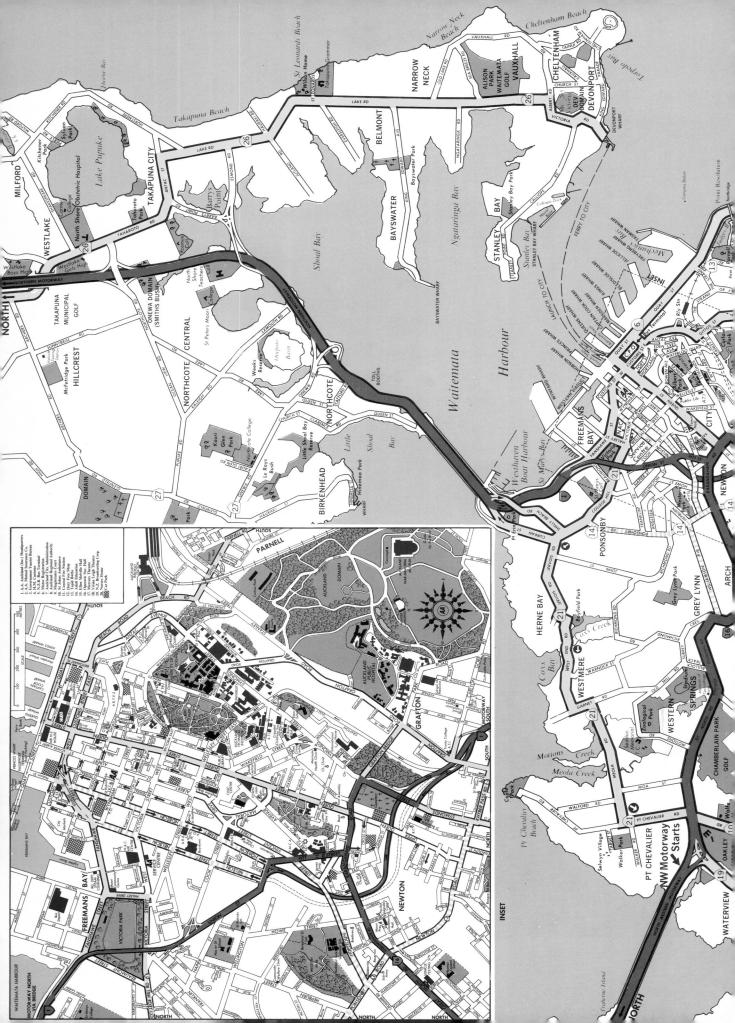

ROUTES IN AND OUT OF

AUCKLAND

SCALE

0   1 km

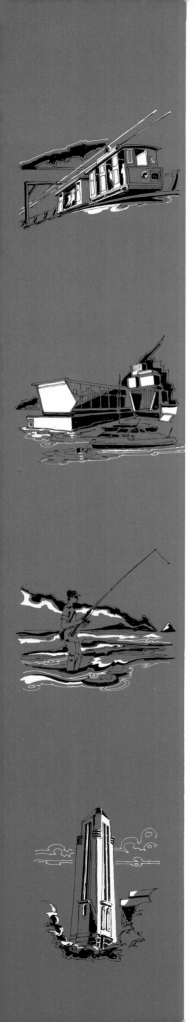

WELLINGTON

Wellington, capital city of New Zealand, is the seat of Government as well as the location of most head offices of national and international organisations, and of agricultural, scientific and industrial bodies, cultural archives and records. It is also the home of representatives of foreign governments and of nations of the Commonwealth. The city is named after the Duke of Wellington.

Wellington's character has been shaped by its role as political capital as much as it has by the demands of its terrain. Government departments and business head offices are crammed on the small fringe of re-claimed land around the harbour. It is interesting to note that when Wellington was first settled, walking around what is now Stewart Dawson's corner was impossible at high tide.

Since the 1950's there has been a proliferation of suburbs on the West Coast, and the Hutt Valley has long been the major industrial area; but the focus of business, social and cultural life has remained firmly in the city centre. The bulk of the population works in the city and because of the topography and limited access routes, the arrival of 20,000 commuters daily is an influx of major proportions. Fast electric trains linking the suburbs to the city, and an improved motorway system, have allieviated some traffic problems.

Because Wellington is the site of a very large number of New Zealand's Head Offices, businessmen from other centres form a major proportion of air-passengers to and from the city. A large number of New Zealand families find at one time or another that business promotion involves being transferred to Wellington.

Wellington's importance as the political centre is reflected in the character of its people. Wellingtonians tend to be more reserved than their northern counterparts, more obviously concerned with the serious side of life. This has its redeeming aspects, such as a strong interest in the arts. Wellington and its people have been saved from self-importance by their location and by the whims of the weather.

Geography

Wellington is located on the southern tip of the North Island, near to the geographical centre of the country; its position makes the city the focus of lines of communications linking the North and South Islands. Port Nicholson, the harbour, is almost totally enclosed with a single, narrow — and often dangerous — entrance in the south-eastern corner. It offers an anchorage safe from the gales which frequently rage through the Cook Strait and is regarded as one of the most attractive harbours in the world.

Except for the flat basin of the Hutt Valley to the north, the harbour is surrounded by steep hills which run roughly from north to south. Because of this, access is difficult. There is only one major outlet to the west, the Ngauranga Gorge, and at the head of the Hutt Valley, the Rimutaka Ranges are a considerable barrier to the Wairarapa Plains to the north.

Running in the same north-south direction is the Wellington Fault, the earthquake zone which in fact created Wellington Harbour. The fault line runs along the western shore of the harbour, an almost straight line which is perfectly visible from the air. Although mild tremors are quite common, the last major upheaval took place in 1855.

Climate

Wellington is justifiably renowned for its winds. Gales with velocities of 103 km/h are recorded on 146 days, and of 96 km/h on 30 days of the year. Gusts have exceeded 160 km/on on occasions. Fine, windless days are rare.

Wellington enjoys 2,010 hours of sunshine each year with a mean summer maximum of 20°C. July is the coldest month with a mean minimum of 5°C. Rainfall is 1206 mm, July and August being the wettest months. Snow is unknown, but hail occurs about 12 times a year and frosts are quite common in the hill suburbs and the Hutt Valley between May and September.

History

Wellington's early years are inextricably linked with Edward Gibbon Wakefield and the New Zealand Company. Captain James Cook was the first European to enter Wellington Harbour in 1773, but it was named Port Nicholson by Captain James Herd who investigated it as a possible site for settlement in 1827. Its modern history really dates from February 1840 when the first settlers sent out from England by the New Zealand Company arrived and settled at Petone, then named Britannia. The first three years were difficult. There was little flat land available for farming, prices were poor and flooding was a serious threat. The establishment of British rule brought troubles over land titles and the Maoris were soon aroused to warfare by the influx of settlers and their demands for land.

In 1865 Wellington became the capital, a move Wellingtonians thought should have been made in 1840 when the seat of Government was shifted from Russell. As roads and railways linked the port with the potentially rich farmland of the Wairarapa, Manawatu and Rangitikei, Wellington began to develop as a business and commercial centre, a growth that was accelerated by its central position and political importance.

Population

Estimated as at April 1 1973: 337,680

Points of Interest

Mt. Victoria (196m) offers a panoramic view of the central city, nearby suburbs, the harbour and a distant view of the Hutt Valley, from the Centennial lookout on the Summit.

Botanic Gardens in Kelburn also offer good views of the city and harbour. Accessible from city centre by cable car.

Museums and Art Galleries

National Museum and National Art Gallery in Buckle Street, on a commanding site formerly called Mount Cook. Museum noted especially for Maori and Polynesian collection. Gallery includes 18th and 19th century European paintings and a New Zealand collection. Hours: Monday to Saturday, 10am to 5pm.; Sunday, 1pm to 5pm. Here is also the War Memorial Carillon containing 49 bells.

Public Library. Located in a modern building in Mercer Street

Alexander Turnbull Library. Perhaps the greatest collection of material including manuscripts on New Zealand and Pacific. Treaty of Waitangi on display. Also collection of early New Zealand paintings. Hours: Monday to Thursday 9am to 5pm; Readers only, 7pm to 9.30 pm.

General Assembly Library in Parliament Grounds. Open to the public only during Parliamentary recess and with local M.P.'s recommendation.

Wellington Harbour Board Museum on Jervois Quay. Many relics of the sea and the harbours development. Hours vary.

Castle Collection of Musical Instruments in Newtown. Private collection of over 300 instruments. View by prior arrangement.

Otari Museum of Native Plants (Wiltons Bush) in Wilton. Comprehensive collection of New Zealand species growing in the open in more than 40 ha of native bush.

Buildings

Parliament Buildings in Molesworth Street. An example of late-colonial pretentiousness completed in 1922. Contrasts with the Gothic ornateness of the General Assembly Library on one side and the modern equivalent, the "Beehive", designed by Sir Basil Spence, on the other. Conducted tours available on request.

Old St. Paul's in Mulgrave Street. Designed by Frederick Thatcher and now administered by New Zealand Historic Places Trust.

Government Buildings on Lambton Quay. Completed in 1876. Reputedly largest wooden building in Southern Hemisphere and second largest in world.

Plimmer House in Boulcott Street. Delightful early colonial contrast with surrounding office blocks. Contains a restaurant.

Thorndon. Part of early Wellington which has now been declared on historic reserve. Bounded by Tinakori Road and Bowen Street.

Railway Station. An imposing structure, railway headquarters for New Zealand.

Gardens and Scenic Reserves

Botanic Gardens. Main entrance off Glenmore Street. Includes Lady Norwood Rose Garden, Begonia House, Carter and Dominion Observatories and much native bush in its 25 hectares.

Zoological Gardens in Newton. Has over 700 animals and 3,500 birds. Hours: Monday to Saturday, 9am to 4.30 pm; Sunday, 10am to 4.30pm.

Central Park on the Brooklyn Road. Sixteen hectares of trees with some gardens.

Charles Plimmer Park on the city side of Mount Victoria

Sports Grounds

Wellington is fortunate in having numerous sports grounds within the city area. The chief of these are the Basin Reserve, Athletic Park and Newtown Park.

Because of the difficult topography, many other parks exist in unexpected places about the hills, and handy to the city.

Most sports are catered for in or not far from Wellington, and for the hardy, yachting and small boating are popular on the harbour. Trampers and climbers find the near by and easily accessible Tatarua and Rimutaka Ranges a convenient venue for their activities.

General

Cable Car off Lambton Quay. Rises to suburb of Kelburn and offers extensive city and harbour views at upper terminus. Little changed from original appearance in early 1890s.

Wakefield's Grave in Bolton Street Cemetery. The grave of Edward Gibbon Wakefield remains in one of the few small corners of historic Bolton Street Cemetery to survive the Wellington motorway. R.J. Seddon, a liberal politician and Prime Minister of New Zealand from 1893-1906 is also buried here.

Massey Memorial on Point Halswell. Marble memorial to William Ferguson Massey, Prime Minister 1912-25. Offers pleasant views of harbour and central city.

Katherine Mansfield Memorial in Thorndon, the area in which some of her most famous short stories were set.

ROUTES IN AND OUT OF
WELLINGTON

SCALE

0

km

HAPPY VALLEY

ISLAND BAY

HAPPY VALLEY RD NORTH

VOGELTOWN

VOGELTOWN PARK

MORNINGTON

MILLS RD

BROOKLYN

OWHIRO RD

BROOKLYN RD

MITCHELLTOWN

HIGHBURY

TAITVILLE

KELBURN

WAIAPU RD

MOANA RD

KAIWHARAWHARA RD

KAROR RD

FAIRLIE TCE

KELBURN PDE

UPLAND RD

DEVON ST

THE TERRACE

RAWHITI

UNIVERSITY

VICTORIA

PARK

SALAMANCA RD

BERHAMPORE

LIARDET ST

FARNHAM ST

BRITOMART ST

MACALISTER PARK

ADELAIDE RD

HUTCHISON RD

WALLACE ST

R.C.

Y.W.C.A.

WEBB ST

WILLIS ST

CUBA ST

WINTER SHOW BLDG

TASMAN ST

ADELAIDE RD

BUCKLE ST

TARANAKI ST

COURTENAY PLACE

WAKEFIELD PARK

MUNICIPAL GOLF

LUXFORD ST

ATHLETIC PARK

RINTOUL ST

RIDDIFORD ST

KEIN ST

WELLINGTON HOSPITAL

GOVT HOUSE

RUGBY ST

BASIN RESERVE

CAMBRIDGE TCE

KENT TCE

PIRIE ST

MT. VICTORIA

ORIENTAL BAY

TARANAKI ST WHARF

OVERSEAS PASSENGER TERMINAL

FREYBERG POOL

ST GERARDS

INSET

MARTIN LUCKIE PARK

MELROSE

ZOOLOGICAL GARDENS

MELROSE PARK

MT ALBERT 177

HORNSEY RD

RUSSELL TERRACE

MANSFIELD ST

ROY ST

NEWTOWN

CONSTABLE ST

DUNCAN TCE

ALEXANDRA RD

WELLINGTON COLLEGE

ALEXANDRA PARK

HATAITAI PARK

BUAHINE ST

MOXHAM AVE

MT VICTORIA TUNNEL

ALEXANDRA RD

MT VICTORIA 196

BYRD MEMORIAL LOOKOUT

RADIO STN

LOOKOUT

HATAITAI

ORIENTAL BAY PARADE

Oriental Bay

ROSENEATH

GRAFTON

Point Jerningham

QUEENS DRIVE

ONEPU RD

BAY RD

KILBIRNIE

EVANS BAY PDE

EVANS BAY RESERVE

KILBIRNE

COUTTS ST

CRAWFORD RD

RONGOTAI RD

SALEK ST

TROY ST

BRIDGE ST

Lyall Bay

LYALL BAY

LYALL PARADE

RONGOTAI

WELLINGTON AIRPORT

FOOT SUBWAY

CALABAR RD

COBHAM DRIVE

Evans Bay

MARINA

PATENT SLIP

BURNHAM WHARF

EVANS BAY PARADE

Greta Point

Snapper Point

Kio Bay

Weka Bay

Balaena Bay

Little Karaka Bay

MIRAMAR WHARF

MAUPUIA RD

SHELLY BAY RD

Shelly Bay

MIRAMAR GOLF

RAUKAWA ST

BROADWAY

MONORGAN RD

CRAWFORDS GREEN

HOBART RD

IRA ST

PARA ST

PARK RD

MIRAMAR

MIRAMAR AVE

POLO

MIRAMAR PARK

MIRAMAR NORTH RD

NATIONAL FILM UNIT

DARLINGTON RD

CENTENNIAL PARK

MT CRAWFORD PRISON

STRATHMORE TCE

STRATHMORE PARK

PANAMA TCE

SEATOUN HEIGHTS

FERRY ST

NEVAY ROAD

SEATOUN WHARF

Worser Bay

KARAKA BAY ROAD

Karaka Bay

KARAKA BAY WHARF

Scorching Bay

SCORCHING BAY

RESERVE

BREAKER BAY RD

RESERVE

DUNDAS ST

CHURCHILL PARK

SEATOUN

Point Dorset

Point Gordon

INSET

TORREN

SMITH

ABEL

WEBB

ARTHUR

WIGAN ST

BUCKLE ST

FREDERICK ST

TAINUI ST

JESSIE

ST

BRETH COLLEGE

CH OF CH

MT CRAWFORD PRISON

Police Training School

KENT

CAMBRIDGE TCE

FENNYTON

LORNE

COLLEGE

PRES

ELIZABETH

GREEK

ADV

QUAKER

FERN

TORY

BROUGHAM

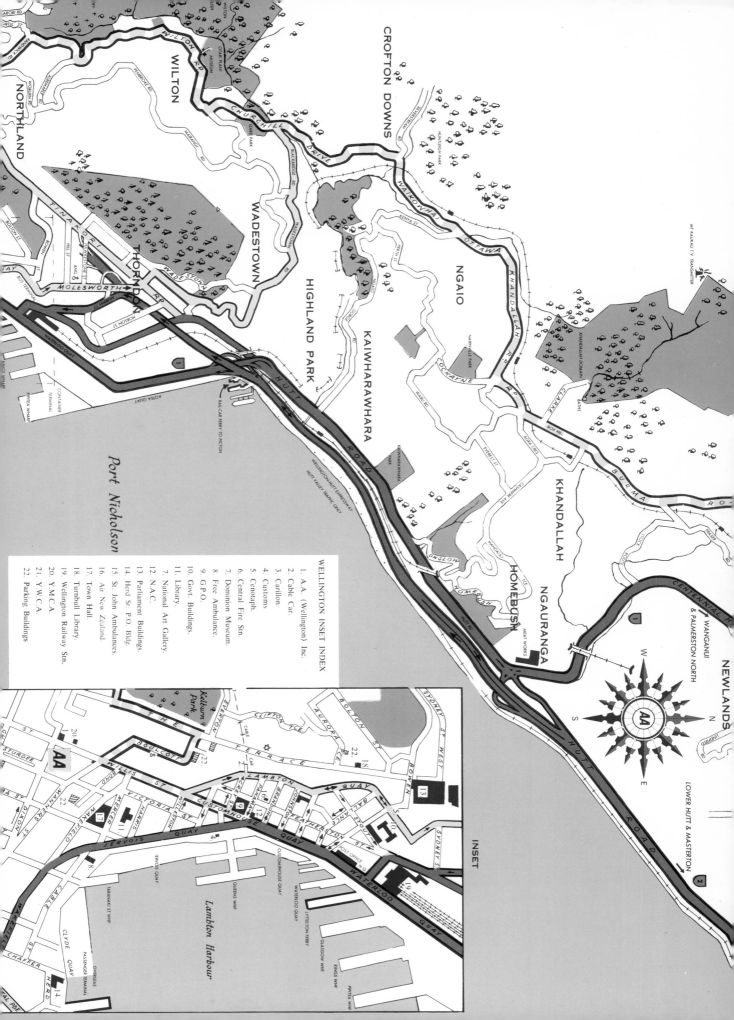

WELLINGTON INSET INDEX

1. A.A. (Wellington) Inc.
2. Cable Car.
3. Carillon.
4. Customs.
5. Cenotaph.
6. Central Fire Stn.
7. Dominion Museum.
8. Free Ambulance.
9. G.P.O.
10. Govt. Buildings.
11. Library.
12. N.A.C.
13. National Art Gallery.
14. Parliament Buildings.
15. St. John Ambulances.
16. Air New Zealand.
17. Town Hall.
18. Turnbull Library.
19. Wellington Railway Stn.
20. Y.M.C.A.
21. Y.W.C.A.
22. Parking Buildings

CHRISTCHURCH

Long regarded as a little English enclave, Christchurch has recently emerged with a new, more progressive image. The opening of the new Town Hall in 1972 and the success of the 1974 Commonwealth Games have drawn attention to other persistent but less obvious aspects of the city's development. Unlike most other South Island urban areas, Christchurch's population is growing at a rate equal to most North Island towns and cities, and it is a national leader in some industries, notably rubber products and specialised engineering, despite its remoteness from major markets to the north. Christchurch is the principal commercial centre of a great primary-producing province. Canterbury is the chief grain and grass-seed exporter of New Zealand and has a large meat and wool industry.

But it is the memory of the genteel facade that tends to remain longest with the visitor to Christchurch: the Law Courts and the Gothic buildings of the old university; the private-school boys with their straw boaters; the Avon lined with willows; and the wide streets converging on the city centre, the Square, with its fine old buildings and Gothic Anglican Cathedral. Perhaps because it has not experienced the same stimuli for rapid growth, Christchurch has been more successful in combining respect for the past with the need for development. The erection of a number of high-rise buildings around Cathedral Square in the last few years shows how precarious this balance between past and present is.

Christchurch also has a well-deserved reputation as a garden city, a reputation which is justified not only by its civic gardens but also by the eager rivalry displayed in best-street and best-garden competitions. It is another illustration, on a smaller scale, of the quiet pride Christchurch residents have in their city and its achievements.

Geography

Christchurch nestles in the lee of Banks Peninsula, the remnant of ancient volcanic activity which juts from the east coast of the South Island. To the west the expanse of the Canterbury Plains stretches to the foothills of the Southern Alps, 48km away. Except for the suburb of Cashmere, Christchurch is built on low, flat ground, a feature which has made construction easy but has brought drainage and sewage problems.

To the west are the Port Hills and the city's port of Lyttelton. Here the extinct crater of an ancient volcano provides an excellent deep-water harbour which has been improved by dredging to cope with the increasing size of modern shipping.

In the first years of settlement Sumner was used as a port, and steamers even ran up the Avon to Christchurch, but the silting of the harbour and the development of larger shipping forced its abandonment. Today Sumner and New Brighton, a few kilometres to the north, are popular summer resorts and dormitory suburbs of Christchurch.

Climate

Christchurch has the lowest rainfall and greatest range of temperature of the four main centres. Rainfall averages 669 mm and is generally evenly spread throughout the year, but the Spring and late Summer are usually drier. Although the mean summer maximum is 22°C, the "nor'wester", a hot dry Fohn wind blowing across the Canterbury Plains, occasionally produces temperatures over 30°C.

Winters are more severe. Between April and October Christchurch averages 36 hard frosts while the mean minimum temperature in July is 1°C. Snow falls occasionally but seldom in any quantity, and hail occurs about six times a year. More annoying are the fogs which sometimes persist throughout the day, aided by the smog produced by Christchurch itself.

History

Christchurch's founding father was John Robert Godley, a devout Anglican landowner. Inspired by Edward Gibbon Wakefield, he and Lord Lyttelton were the driving force behind the Canterbury Association formed in 1848 to organise a Church of England settlement in New Zealand. The intention was to reproduce an ideal English society, carefully graded from labourers to gentry. The ideal society was soon altered by the realities of colonial life, but enough of the original vision remained to stamp Christchurch's character indelibly.

An early handicap was the barrier of the Port Hills which hampered movement between harbour and city. In 1867 a tunnel was completed under the Port Hills, the first major engineering achievement in the colony. Christchurch also had the first railway in New Zealand, a 6.5 km line opened in 1863.

As the rich hinterland of the Canterbury Plains was developed so Christchurch grew too. Progress was remarkably free from the cycle of boom and bust that afflicted most other areas of New Zealand. This was due partly to the careful planning of the original settlement and the leadership that came with the first settlers, but more important was the ease with which the Canterbury Plains were settled, and the diversity of stock and crops it could support.

Until the turn of the century, Christchurch remained largely a service town, a role that promoted close understanding between town and country. In the 20th century secondary industry grew in importance, and Christchurch now rates statistically as the second largest industrial area in New Zealand.

The city was named by Godley for Christ's Church, Oxford.

Population

Estimated as at April 1 1973: 313,210

Points of Interest

City Views

Hackthorne Road. From Hackthorne Road as it climbs through Cashmere to Victoria Park are interesting views of the city, beautiful at night.

Summit Road from Evans Pass to Gebbies Pass offers varied views of the City, plains and Southern Alps throughout its 38.5 km.

Cathedral Spire. The climb up the spiral stone staircase to the Cathedral Spire is worth the effort, for the view of the city from the top.

Museums and Art Galleries

Canterbury Museum in Rolleston Street. Original building designed by noted early architect, B.W. Mountford, and opened in 1870. Special features include the Stead collection of birds; a fine reconstruction of a colonial Christchurch Street, and relics of Antarctic expeditions.

Robert McDougall Art Gallery in Rolleston Street. Good 19th century New Zealand collection, especially related to Canterbury.

Canterbury Society of Arts Gallery in Gloucester Street. Displays by members and touring artists.

Ferrymead, Bridle Path Road. Museum of Science and Industry. Includes rides on steam and electric trams.

Buildings

Christchurch Cathedral in Cathedral Square. Magnificent Gothic structure, designed by Sir Gilbert Scott. Begun in 1864 and finally completed in 1901. The Cathedral has a wide reputation for its achievements in Choral music.

Town Hall, Victoria Street. A magnificent complex of new buildings, comprising conference rooms, a large auditorium, exhibition halls and restaurant. Modern architecture in attractive surroundings beside the Avon River.

Provincial Council Chambers. Picturesque building in Rolleston Avenue. Designed by B.W. Mountford with wooden portion built in 1859, stone section in 1865. Used until the abolition of Provincial Government in 1876. Open weekdays 9 am to 4 pm.

Sign of the Takahe and sister buildings **Sign of the Kiwi** and derelict **Sign of the Tui**, on Summit Road. Part of Harry Ell's vision of Road Houses at easy stages over the Port Hills road to Akoroa.

The Sign of the Takahe is the most complete. It is faithfully Gothic in character and design, and was carefully made by craftsmen.

Canterbury University, Ilam Road, Ilam. Beautiful new buildings in a variety of architectural styles, set in spacious grounds. The University moved here in the 1960's forced out of the city by overcrowding. The historic old buildings in Rolleston Avenue have been preserved as a cultural centre.

Roman Catholic Cathedral in Barbadoes Street. Designed by F.W. Petre with fitting Roman influence.

Cob Cottage on Sumner Road. Now rare historic cottage constructed of cobs.

Parks, Gardens and Reserves

Queen Elizabeth II Park. This imposing complex of a design rare in the world of sport was built in two years for the Xth British Commonwealth Games in 1974. It is designed for a multitude of sports, especially swimming and track events, and has some of the best sporting facilities in New Zealand.

Avon River. This small river winds around the centre of the city, its banks lined with beautiful lawns and trees.

Hagley Park in city centre. Over 1800 hectares set aside by the pioneers, including playing fields, a lake, golf course and Botanic Gardens. This huge park adds particular beauty and openness to central Christchurch.

Botanic Gardens. Established in 1863 adjacent to Hagley Park. Fine collection of tropical plants, cacti and succulents. Excellent rose and bulb section, with cherry trees, rhododendrons and other exotic plants.

Victoria Park on Port Hills, access via Hackthorne Road. Views of city and 72 hectares of trees and scrubs.

Deans Bush at Riccarton. 6.4 hectares of land owned by first settlers in district.

Latimer Square and Cranmer Square. Small green areas in the city, named for the martyred Bishops.

Floral Clock in Victoria Street is an excellent example of artistic gardening.

General

Bridge of Remembrance over the Avon at Cashel Street. Memorial to two world wars.

Totem Pole in Harper Avenue. Given to the city by the state of Oregon, U.S.A., in recognition of the city's hospitality to American personnel stationed in Christchurch for operation "Deep Freeze".

Christchurch International Airport. A large modern airport already being extended to meet the demands of increasing traffic. Adjacent is the American "Deep Freeze" operations base, for the servicing of the U.S.A.'s McMurdo Base in Antarctica.

Memorial Cross, Cathedral Square, is a memorial to those who served in two world wars.

Bowker Fountain, Victoria Square, plays nightly. It is of interest that this was the first coloured fountain in Australasia. There is a fine statue of Captain James Cook.

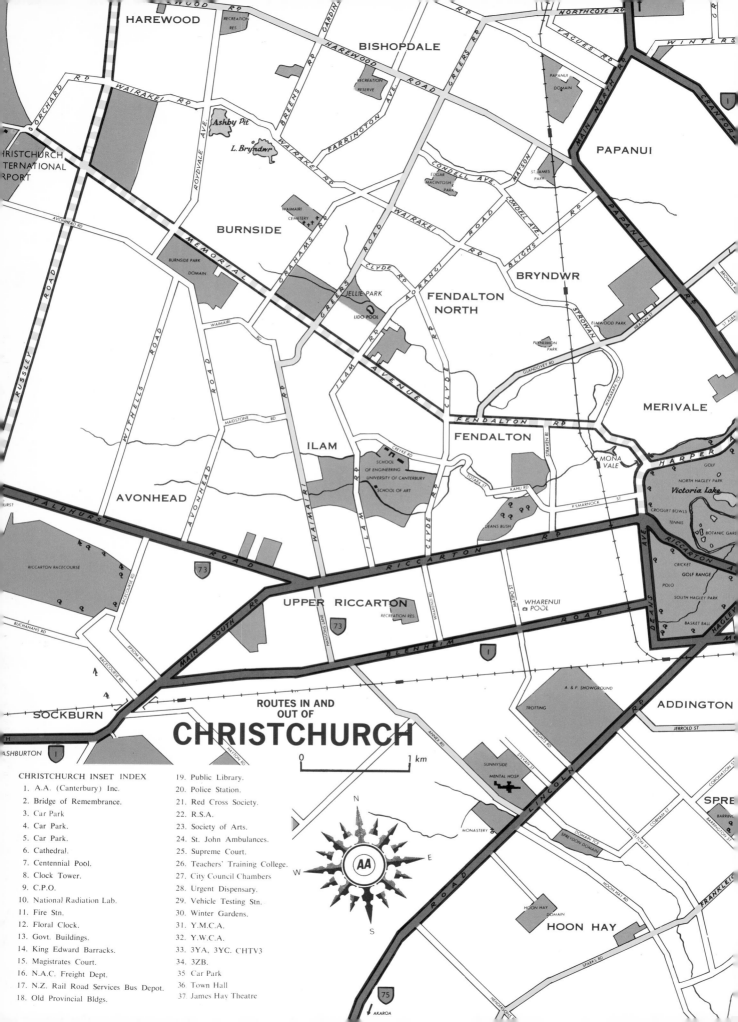

ROUTES IN AND
OUT OF
CHRISTCHURCH

0   1 km

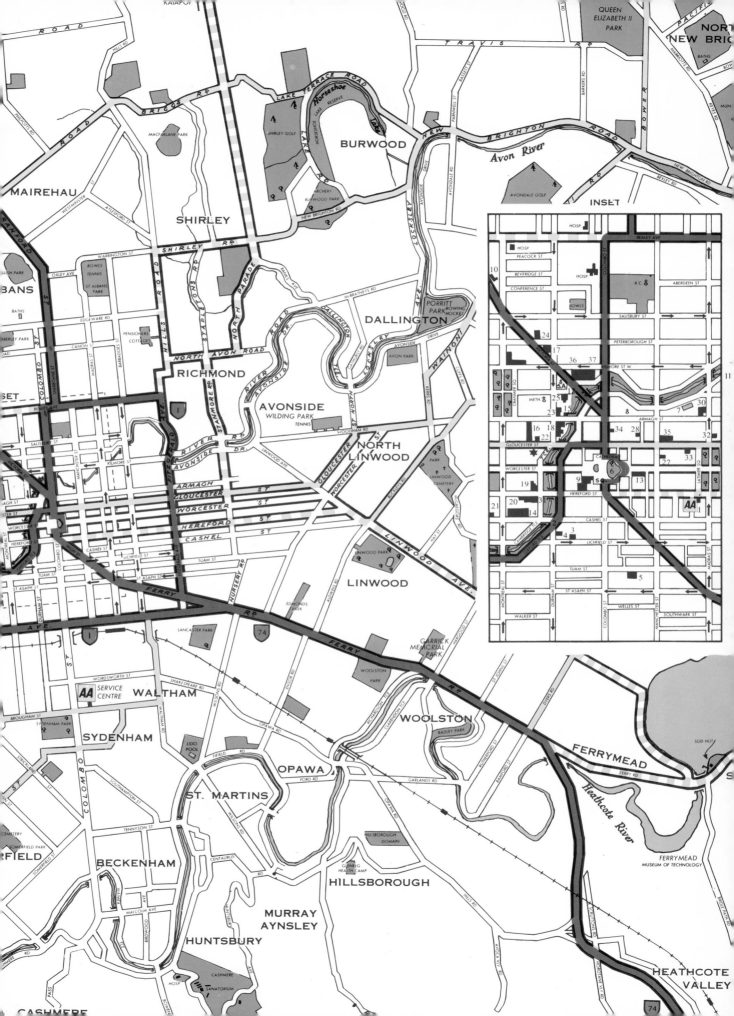

DUNEDIN

In retrospect it seems almost appropriate that New Zealand's Scottish founders should have chosen Otago as their new home. There is something in Dunedin's rugged location and the more rigorous climate that fits in well with the Scottish character.

Yet it is a mistake to regard Dunedin as no more than the "Edinburgh of the South", or to see it as a museum of past achievements. Although overall growth has suffered with the shift of population north, Dunedin is acknowledged to be financially the soundest community in New Zealand. Industry has always been marked by vigour, enterprise and variety, and the city is a nationally important producer of foodstuffs, textiles and transport equipment. Wool and frozen meat form the major export business of the port.

Dunedin is regarded as the "University City" of New Zealand. There is an unusually close relationship between the community and its University, and several of the courses offered are still unavailable elsewhere.

With a slower way of life, the people of Dunedin appear to have more time and energy to devote to cultural pursuits, and achievements in the arts are impressive. Equally memorable is the friendliness of the people, a characteristic that perhaps best sums up Dunedin's heritage from its Scots and pioneer past.

Geography

Dunedin lies in a setting of great natural beauty at the head of Otago Harbour, an inlet running 24km in a southwesterly direction and creating Otago Peninsula on its seaward side. Approximately half way along the Harbour is Port Chalmers, Dunedin's second port.

The industrial portion of the city is built on flat land, much of it reclaimed from the harbour; the central business district is on the lowest slopes of the hills, and behind, the land rises steeply into hill suburbs. Behind the city is a sweep of higher hills running from south to north east, the highest points being Saddle Hill, Flagstaff and Mt. Cargill. Further west lies the flat fertile expanse of the Taieri Plains, and the great plateau of grazing land that has been the source of much Otago wealth.

Climate

With its more southerly position, Dunedin's weather is cooler. The warmest months, January and February, have a mean maximum of 19°C, while July is the coldest month with a mean minimum of 3°C. The city usually records an average of 1,730 hours of sunshine a year.

Rainfall is spread evenly throughout the year with a relatively low annual average of 790mm. Snow falls about twice a year, but this is usually only a few mm and some winters are free of snow. Hard frosts are recorded on 12 days of the year, and hail on about 10 days. The city experiences frequent sudden changes of weather as cold fronts bring showery south-westerly conditions. North-easterly winds are also common and sometimes bring fog and mist. North-westerly weather is hot and dry but not as drastically so as in Canterbury.

History

Shore whalers were the first European settlers on the site of Dunedin. They knew the place by its Maori name, Otakou, which they corrupted to Otago and applied to the harbour area, and eventually to the whole province.

Dunedin city had its origins in organised settlements. In Scotland in the 1840's, leading members of the Free Church of Scotland — notably the Rev. Thomas Burns (nephew of Robert Burns) and Captain William Cargill — planned a church settlement in the South Island of New Zealand, and employed surveyor Frederick Tuckett to find a suitable location.

Two small sailing ships — the "John Wickliffe" and "Philip Laing" — brought the first colonists to Dunedin in early 1848 — 344 Scots with a strong determination to found a society based on solid principles of religion, education and industry. Although the city is now much more cosmopolitan and varied, the original purpose of its founders has left an indefible mark on its character and institutions.

The greatest impetus for growth was the discovery of gold in Central Otago by Gabriel Read in 1861; in two years the population rose from 2,500 to 10,000 and Dunedin, previously small and struggling, became very suddenly a busy, wealthy commercial centre. By the 1880's it was the largest and most industrialised city in New Zealand — a preponderance it lost only when the North Island shook off the burden of the Maori Wars and began to surge ahead. The northward drift of population began in 1900 and has not ceased.

In 1882 the first refrigerated cargo of New Zealand meat left Port Chalmers for England on the Ship "Dunedin"; the freezing works at Burnside was the first in the world.

Since 1907 the city has drawn its electricity supply from the Waipori Power Stations, privately owned by the Dunedin City Council. The Council also owns 6880 hectares of land for forestry, and tree planted 40 years ago are the beginning of a log trade through the Port of Otago.

Dunedin attained city status in 1865.

Population

Estimated as at April 1973: 118,970

Points of Interest

City Views

Rotary Park on the Peninsula High Road, via Anderson's Bay. A lookout and parking area provided extensive views of the city, and of the harbour right to Taiaroa Heads. The city view is particularly attractive at night.

Lawyers Head at the northern end of St. Kilda Beach, via John Wilson Memorial Drive, has attractive views of the beaches and coastline.

Signal Hill Access from Opoho. Panoramic view of City and harbour. Site of New Zealand Centennial memorial. The sculptures are by F.A. Shurrock of Christchurch.

Thomas Bracken Lookout in Lovelock Avenue. Panoramic views of city.

Unity Park, Eglinton Road. Excellent view of the city and harbour, Jaycees have put a locality map here.

Town Belt A strip of lovely native bush, small parks and exotic trees, that divides downtown Dunedin from the main residential suburbs. From Queens Drive, winding along its length, are attractive glimpses of the harbour and city. Prospect Park at the northern end has parking space and unobstructed views.

Museums and Art Galleries

Otago Early Settlers' Museum in Lower High Street. Many old paintings, photographs and detailed records of Otago families and history. There is a fine collection of old horse-drawn carriages. Hours 9 a.m. to 4.30 p.m. Monday to Friday.

Otago Museum, Great King Street — Cumberland Street, (1877) has in addition to the usual exhibits, a magnificent collection of New Zealand and Pacific ethnological material in the Skinner Hall of Polynesia; and a notable collection of Egyptian antiquities. Art exhibitions are frequently staged in the Museum foyer. Hours: weekdays summer, 10am. to 5pm., winter 10am. to 4.30pm.; Sunday 2pm. to 4pm.

Hocken Library in Otago Museum. An impressive collection of books, newspapers, maps, letters, manuscripts, pictures and portraits relating to New Zealand and Pacific History. Given to the city in 1897 by Thomas Hocken, Surgeon, Bibliographer and Collector.

Dunedin Art Gallery in Logan Park at the end of Anzac Avenue. Some notable New Zealand paintings and a fine British and European collection, particularly of the early English Water-colour School. Hours: Monday to Friday, 10am to 4.30pm; Saturday 10am to 5pm; Sunday 2pm to 5pm; holidays 10am to 6pm.

Portobello Aquarium Marine aquarium and Biological station, on the Portobello Peninsula, operated by the Otago University. A comprehensive, fascinating collection of marine life. Hours: 9am to 5pm daily.

Buildings

Olveston in Royal Terrace. A unique Jacobean-style mansion in large grounds, bequeathed to the city in 1966 by Miss Dorothy Theomin. It is beautifully furnished and provides insight into an age of elegance. Guided tours Monday to Saturday, 9.30am., 10.45am., 1.30pm., 2.45pm., 4.00pm., Sunday, 1.30pm., 2.45pm., 4.00pm.

First Church in Moray Place (Presbyterian) Magnificent Gothic church with an aesthetically tapering spire, designed by R.A. Lawson and built in 1873. The Cameron Centre in the church grounds, is the Presbyterian church's headquarters for social services in Otago.

Otago University Leith Street, Castle Street, Union Street. The original buildings are of stone, grouped around a quadrangle, but the University has since spread over a large area. As well as the faculties of Arts, Science, Commerce, and Law, special schools provide degree courses in Medicine, Dentistry, Physical Education, Home Science, Divinity, and Technology (incorporating the Otago School of Mines).

Otago Boys' High School, established 1883. Beautiful stone building with a tower, in attractive park surroundings. Designed by R.A. Lawson. New buildings have been especially designed to blend with the old. Campbell House nearby is the school's boarding house.

Larnach Castle on Otago Peninsula, off Highcliff Road. A grand "Scottish Baronial" Mansion built in the 1870's by the Hon. W.M.J. Larnach — the most expensive ever built in the Southern Hemisphere. Open daily 9am to dusk.

Gardens and Scenic Reserves

Botanic Gardens at north end of Great King Street. Interesting and varied, with several famous collections, covering a considerable acreage of hillside. There is a small zoo near the entrance.

The Octagon. A small green area in the centre of the city, with the "Star" Musical fountain, and a statue of Robert Burns. Around it are the Town Hall, St. Paul's Cathedral, The Athenaeum Library, and the Regent Theatre, which is to become Dunedin's Civic Theatre.

Glenfalloch Woodland Garden near Macandrew Bay on Portobello Road. A lovely showplace of trees, shrubs and gardens, now owned by the Peninsula Trust and open to the public.

Queen's Gardens. An open area of grass and large old trees around the Cenotaph. There is a fine statue of Queen Victoria.

General

St. Clair and St. Kilda on the south-eastern outskirts of the city, have popular bathing beaches. There are warm salt water baths at the southern end of St. Clair beach.

Moana Pool. An Olympic-sized swimming pool, with all facilities, in pleasant surroundings on the edge of the Town Belt. Hours Monday to Saturday, 7am to 5pm., 7.30pm to 10.30pm, Sundays 10am to 5pm.

Captain William Cargill monument in Princes Street. A neo-Gothic Victoria structure commemorating Dunedin's founder.

Beverly Begg Observatory in Ross Street, with picnic area nearby. The Observatory on Signal Hill belongs to Otago University.

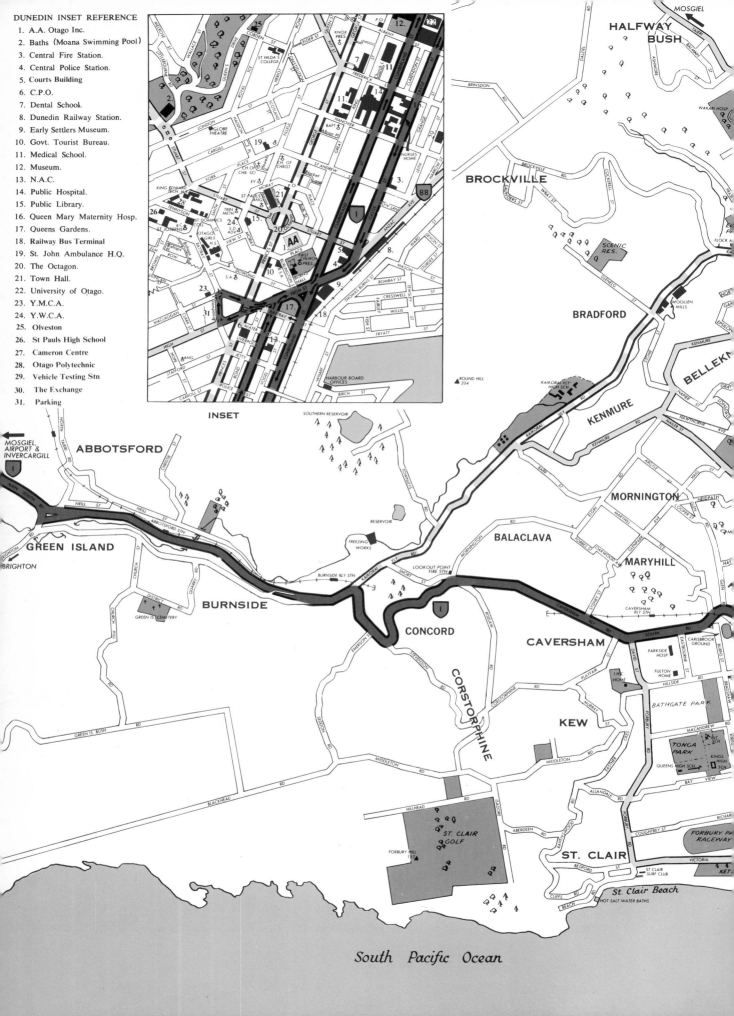

DUNEDIN INSET REFERENCE

1. A.A. Otago Inc.
2. Baths (Moana Swimming Pool)
3. Central Fire Station.
4. Central Police Station.
5. Courts Building
6. C.P.O.
7. Dental School.
8. Dunedin Railway Station.
9. Early Settlers Museum.
10. Govt. Tourist Bureau.
11. Medical School.
12. Museum.
13. N.A.C.
14. Public Hospital.
15. Public Library.
16. Queen Mary Maternity Hosp.
17. Queens Gardens.
18. Railway Bus Terminal
19. St. John Ambulance H.Q.
20. The Octagon.
21. Town Hall.
22. University of Otago.
23. Y.M.C.A.
24. Y.W.C.A.
25. Olveston
26. St Pauls High School
27. Cameron Centre
28. Otago Polytechnic
29. Vehicle Testing Stn
30. The Exchange
31. Parking

INSET

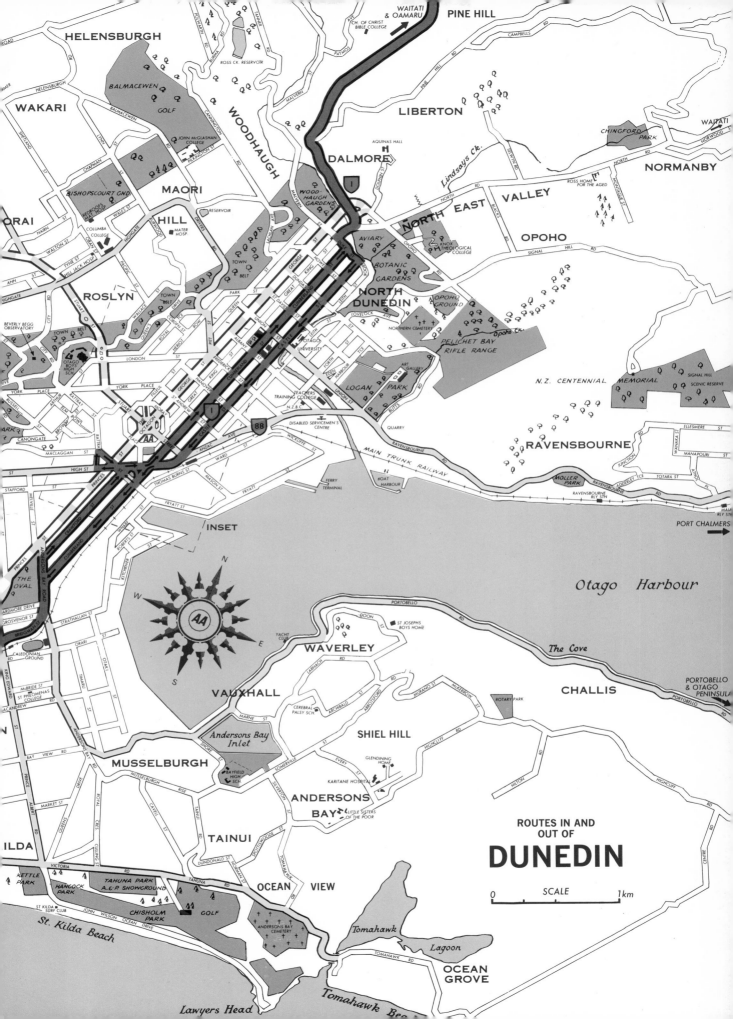

NORTH ISLAND

NORTHLAND

Northland is the long thin arm of land stretching out to the north-west from the main bulk of the North Island. Almost 320 km long, its coastline is so deeply indented, especially on its eastern side, that the region occupies only five per cent of New Zealand's land area.

The west coast from Manukau Heads to Reef Point forms an almost straight line broken principally by Kaipara Harbour, about 70 km north of Manukau, and Hokianga Harbour, approximately 120 km further north. At Reef Pt. the land curves eastward before resuming its north-west thrust along Ninety Mile Beach to Cape Maria van Diemen and Cape Reinga. In contrast, much of the east coast is an intricate mosaic of harbours, headlands, inlets and islands, especially around the aptly named Bay of Islands.

The landscape is as complex as the coastline, its geological contents heightened by man's influence. Largely because of the variety of soil types, the pattern of land use forms a giant patchwork: pastures interlock with desolate gumfields, and the ordered charm of vineyards and citrus orchards contrasts with the grandeur of kauri forests.

History

Northland was the cradle of European settlement and perhaps also the site of the first Polynesian homes. By the early 19th century, Northland held the largest, and possibly most vigorous, population of the Maori people.

The first Europeans were a motley collection composed mainly of whalers, sealers, traders and runaway convicts. Despite this, conflict between Maori and European was limited until 1845 when Hone Heke led a rebellion against the newly established colonial government. And there was a redeeming aspect for Northland was the ground for the first missionary work in New Zealand.

After the heady days of the 1840s, Northland's development slowed, and even gum digging and the milling of the native bush only partially offset this trend. It was not until the 1950s that Northland really experienced a resurgence with the improvement of roading and research into soil mineral deficiencies.

SOUTH AUCKLAND-WAIKATO

South of the steadily encroaching urban sprawl of Auckland lies some of the richest farm land in New Zealand. South Auckland and the Waikato form a continuum, both geographically and in character. The northern boundaries are formed by the Manukau Harbour, Auckland's southern suburbs and the Firth of Thames; to the west is the Tasman Sea. The eastern margins are marked by the Coromandel and Kaimai ranges and the southern limits are set by the rougher hill area of the King Country.

The land within these limits is generally low-lying, rising at the most into rolling hills generally under 300m. To the north of Hamilton and on the Hauraki Plains to the north-west the land is frequently swampy, a legacy from the wandering of the Waikato River which at one stage discharged into the Firth of Thames. Although considerable areas of swamp remain, most notably around Motumaoho, Rukuhia and Moanatuatua, and at Piako on the Hauraki Plains, an even greater area has been drained and converted into highly-productive farms.

History

Perhaps no other part of New Zealand has been tamed as thoroughly as South Auckland and the Waikato. Its present gentleness contrasts with its past. The Waikato was the centre of the Maori King Movement, a Maori land league formed in the 1850s to protect Maori land from the demands of Pakeha settlers. The result was war in the 1860s and victory for the colonists largely through superior weapons and the lack of unity among the various Maori factions. Defeat brought large-scale land appropriations and the rapid expansion of European settlements. For the first soldier-settlers conditions were often difficult, but the development of refrigeration in the late 19th century laid the basis for the present prosperity.

Few signs remain of the battles of more than a century ago. At Rangiriri between the Auckland-Hamilton highway and Lake Waikare, are old fortifications, site of the largest and probably crucial battle of the Waikato War. Just south-east of Te Awamutu is the site of the last engagement at Orakau, now a road-side picnic area.

COROMANDEL

The Coromandel Peninsula juts out into the Pacific Ocean, separating the Hauraki Gulf and Firth of Thames from the Bay of Plenty. In many ways it resembles an underdeveloped twin to Northland, a parallel supported by its warm climate, its many bays and inlets and the richness of its history. Unlike Northland, it has not shown a resurgence of growth in recent years.

It is a region of steep, broken hills, with flat land confined to valley floors. Apart from farming on these flats, most of the Peninsula is covered with regenerating forest.

History

The Coromandel Peninsula was an early centre of European activity; in 1794 the first load of spars was cut from the forest cloaking the hills. By the 1820s timber milling was a growing activity but in 1851 an even greater stimulus was given with the discovery of gold by Charles Ring. This rush was shortlived, but a bigger and more lasting one began in the 1860s with a further big find in the 1880s. After the first easy pickings, mining was largely concentrated in larger companies which could afford the crushing equipment needed, but gold, gumdigging and timber, especially kauri, provided employment for a sizable population. As these three resources were exhausted around the turn of the century, most of the population shifted away in search of other opportunities.

BAY OF PLENTY — VOLCANIC PLATEAU

Named by Captain Cook during his first voyage in 1769-70, the Bay of Plenty has justified its title with its growth since World War II. This prosperity has been shared by the Volcanic Plateau further south, for the two regions are closely linked economically and geographically.

The western boundary is formed by the bulk of the Kaimai and Mamaku Ranges. The line continues south-east to include Lake Taupo and the central mountains Tongariro, Ruapehu and Ngauruhoe before curving to the east around Waiouru. Hill country also marks the eastern margin: from the rugged Urewera district to the east of Opotiki the boundary runs through the hill country separating Taupo from Hawke's Bay to the Kaimanawa Mountains.

A belt of volcanic activity runs through the centre of the region. Reaching from White Island, offshore from Whakatane, to the mountains in the centre of the North Island, it is a reminder of the natural forces which formed the landscape. There are few extensive areas of flat land; most is confined to the coastal region, the margins of the Waikato and Rangitaiki rivers, and portions of the Volcanic Plateau. Elsewhere the terrain varies from rolling to rugged hill country.

The northern boundary of the region is firmly defined by the Pacific Ocean.

History

Like most of the northern part of the North Island, the Bay of Plenty supported a sizable and vigorous Maori population before the coming of the European. The first Pakeha settlers, traders in the 1820s, found themselves embroiled in the wars between the tribes, that marked that decade. War again returned to the area in the 1860s. In the fighting between Maori and European, the Arawa people around Rotorua supported the British, but the Tauranga Maoris allied with the Kingites and won a notable victory over Imperial troops at Gate Pa in 1864.

Despite the persistant threats of guerilla raids by Te Kooti and his followers until the mid-1870s, European expansion began in the 1860s as tourists from New Zealand and throughout the world were attracted to the termal wonders of the Rotorua region. The destructive potential of these wonders was demonstrated in 1886 when Tarawera erupted, killing approximately 150 people and scattering ash over hundreds of square kilometres.

The spread of farming was slower and was confined virtually to the area north of Rotorua until the 1930s when the discovery of cobalt deficiency in the volcanic soils opened the area for farming. Until then the area south to Taupo and beyond had been regarded as "wasteland", suitable only for the pinus radiata forest planted extensively in the 1920s and 1930s.

EAST COAST — UREWERA

The north-eastern "corner" of the North Island has been a forgotten region for much of the 20th century. A rugged bastion thrust out into the Pacific Ocean, it has only recently begun to emerge from its economic stagnation. The region begins just to the east of Opotiki where the coastline changes its gentle south-east gradient and turns to the north-west. This north-west angle is continued inland by the Urewera Country as far as Lake Waikaremoana. The northern and eastern boundaries are set by the coastline, the southern margin by the hill country to the north of Wairoa in Hawke's Bay.

The East Coast and Urewera is a region of steep, broken hills falling away from a central spine, the Raukumara Range. Almost 80 per cent of the land is classed as moderately steep to steep, only 9 per cent as flat. Much of the western portion of the region is still cloaked in bush and has been incorporated in the Urewera National Park. With its area of 199,530 ha it is the largest national park in New Zealand and includes in its south-eastern corner beautiful lake Waikaremoana.

History

The majority of the Maori inhabitants of the area settled on the coastal fringes, living on the wealth of the sea and the resources of the forest fringing the coast. Only the Tuhoe, the "children of the Mist", made their home in the rain-shrouded, forested hills of the Urewera.

It was on the East Coast that James Cook made his first landfall on October 7 1769.

KING COUNTRY

The division between the King Country and the Waikato is historical rather than physical. Here the leaders of the King Movement retreated after their defeats in the Waikato, Taranaki and elsewhere, and for almost 30 years the area remained closed to the Pakeha.

The Tasman Sea sets the western boundary, while the eastern area is separated from the Volcanic Plateau by the Rangitoto and Hauhungaroa ranges. The Raetihi and Ohakune districts are on the southern border with Wanganui.

History

European penetration into the tribal homeland of the Ngati Maniapoto was limited before the wars of the 1860s. In 1864 Tawhiao, the Maori King, and many of his followers retreated here, abandoning their land to the Europeans. Although the Ngati Maniapoto had taken a prominent part in the Taranaki and Waikato fighting, none of their land was confiscated, perhaps because of its apparent unsuitability for farming. Here other "rebel" leaders and their followers congregated as the fighting flared and was suppressed in other parts of the North Island.

In 1884 a route for the Main Trunk line was surveyed through the region, and it began to open up to Europeans. By 1887 the railways had been extended from Te Awamutu to Te Kuiti, but it was not until 1908 that the Main Trunk line was completed. Although farming spread in the Otorohanga and Te Kuiti districts, and into the hill area north of Wanganui, most of the European settlement was close to the railway, and timber milling was an almost equally important industry.

TARANAKI

The Taranaki landscape is dominated by Mount Egmont; few parts of the region are without a view of its near-perfect cone. From the mountain itself the land falls away into rolling country under 300 m. To the north, west and south this easy country extends to the sea; only to the east does it rise into higher, more difficult hill country on the King Country border. A feature of the region is the number of small rivers and streams that radiate in all directions from Mount Egmont.

History

Populated by the descendents of four canoes, Taranaki marked the southern limit of intensive Maori settlement, especially around the coast and up the river valleys.

Abel Tasman sailed up this coast in 1642, but it was James Cook who gave the mountain at Taranaki its present name. In the 1820s traders arrived and, soon after them, northern Maoris armed with muskets. By the end of the 1830s, when Colonel William Wakefield arrived to buy land for European settlement, he found the land virtually unpopulated, but the original inhabitants began to return after 1840 and tension between them and the land-hungry settlers grew. It erupted into war in 1860 and for the rest of the decade fighting continued with the struggle taken up in its later phase by the more fanatical Hauhaus. In 1870 the conflict took a new turning under the Maori leader Te Whiti who introduced a form of passive resistance very similar to that made famous by Gandhi in India.

WANGANUI

The Wanganui region has been shaped by the river which has given it its name. It is a narrow strip of land running inland from near Patea in the west and the Whangaehu River in the east to just south of Raetihi. Most of the land is rugged and mountainous, and the river has cut a deep course throughout most of its length. Only the last 16 km broadens into a narrow plain barely 4 km wide at its lower reaches.

History

To the Maori, rivers were a favoured means of communication, and the Wanganui was one of the great highways into the interior of the country. The first Europeans to venture on the river in 1831 had little time to appreciate it as all except one were killed because of their connection with the shrunken head trade. This event was long forgotten when the tireless Colonel Wakefield arrived and bought 16,200 hectares for a few hundred dollars' worth of goods. Settlers poured into the new town, then known as Petre, but Maori dissatisfaction at the price they had received for their land grew. Led by Topine te Mamaku, who had assisted Te Rangihaeata in the Hutt Valley in 1845-6, the anti-sale faction opened hostilities in 1847, but were soon deterred by the presence of 800 Imperial troops and sailors — and the even more effective threat from pro-European Maoris in the area.

Growth was slow, hampered by the lack of easily-cleared land and the difficulty of overland communications. The establishment of road and rail to the north-west and south-west brough expansion as Wanganui became the port for southern Taranaki and also the western portion of the Rangitikei. The river trade was important too. By the late 1880s shallow-draught steamers were making regular trips as far as Taumarunui despite the hazards of numerous rapids. This ceased in 1934 with the deterioration of the river and the improvement of roads.

RANGITIKEI-MANAWATU

Over 9,000 square kilometres in area, the region's topography varies from coastal dune areas to steep, high-country hills. The western boundary follows the line of the Whangaehu River inland from the coast. The southern extremities of the Kaimanawa Range north of Taihape mark the northern boundary. Hills also form the eastern margins of the region; the Ruahine and Tararua ranges separate Rangitikei and Manawatu from Hawke's Bay and the Wairarapa respectively. To the south the region narrows as the sea curves in towards the Tararuas until, at the boundary with Wellington south of Waikanae, it is little more than a strip of coastal flats between sea and hills.

Sand dunes extend inland for up to 16 km from the coast. Much of this dune area is farmed and erosion is a serious problem. Behind the dunes are the alluvial plains built up by the two major rivers the area is named after. The northern part is 16 to 24 km wide and it tapers towards the southern apex. To the north and east the land rises to low hills and then higher hill country on the outskirts of the region.

Between the Ruahines and Tararuas is the Manawatu Gorge, a narrow opening which provides one of the major access routes to the east.

History

The Manawatu-Rangitikei was an early area of settlement. The first Polynesian migrants hunted here over 1,000 years ago and burned large areas of forest. Settlements along the river were numerous in the Classical Maori period also. A new phase began with the arrival of the redoubtable Te Rauparaha in the early 1820s; he established himself on the fortress island of Kapiti off Waikanae.

The first years were difficult as the heavy bush and transport difficulties slowed development, but within a few decades the bush had been transformed into farmland. Development of the northern portion around Taihape was later and less rapid, though pastoralists spread from the Hawke's Bay into the open, tussock area in the far northern portion. A major stimulus to European expansion here was the construction of the Main Trunk line completed in 1908.

HAWKE BAY

Hawke Bay's northern boundary is formed by the lower hill country just to the north of Wairoa. To the west the broken high country of the Kaweka and Ruahine ranges separate the region from the Volcanic Plateau and the Manawatu-Rangitikei districts. The southern boundary is a horizontal line stretching from the Manawatu Gorge to the east coast.

Within these limits, approximately 193 km in length and less than 65 km wide, is some of the most productive farmland in New Zealand, particularly the Heretaunga Plains and their extension, the Ruataniwha-Takapau Plains. To the east and north the land rises fairly gradually into the ranges which form one of the region's boundaries.

History

The first European settlers in the district found much of the northern and central part covered by scrub and tussock, a legacy perhaps from distant moa-hunter days. The largest areas of open land naturally attracted early settlers, and by 1849 J.H. Northwood and H.S. Tiffen had established a 16,200 hectare sheep run. Others soon followed and by 1874 most of the area north of Takapau was taken up in runs.

South of Takapau was the dense forest of the Seventy Mile Bush. This portion of the province was largely ignored until the 1870s when 5,000 Scandinavian families were settled in the area which extended south into the Wairarapa. Their story was one of hardship and great effort, but by the turn of the century most of the bush had been cleared.

The Seventy Mile Bush formed a barrier to the Wairarapa and Manawatu, but despite this Hawke's Bay's growth was steady, and it had become a prosperous province within a few decades of its founding. One possible reason for its early success was the virtual absence of the fighting between Maori and European which wracked other areas further north in the 1860s. By 1878 Napier was the seventh largest urban area in the country and by the 1890s, the decade in which it was linked with Wellington by road and rail, it was the fifth largest.

WAIRARAPA-WELLINGTON

Wellington and Wairarapa are separate regions geographically and to a lesser extent historically, but are closely linked economically. The Wairarapa occupies the eastern portion of the southern tip of the North Island. Beginning just south of Waipukurau, its eastern and southern boundaries are formed by the sea, its western limits by the Tararua and Rimutaka ranges.

Wairarapa's northern region is a continuation of the Hawke's Bay plains which gradually narrow between the Tararua and Puketoi ranges. South of Mount Bruce, the native bird reserve, the land opens out again into the Wairarapa Plains formed largely by the Ruamahanga River. The southern part of the Plains, between Lakes Ferry and Onoke, is swampy and lowlying. To the east is a region of hill country, much of it steep and eroding, and up to 48 km wide.

The Wellington region is a narrow strip of land between the west coast and the Tararua Range. Flat and even rolling country is minimal, and most of what there is has been swallowed by the expanding cities and dormitory suburbs.

History

The first European settlers who arrived at Petone on Port Nicholson in 1840, found the local Maori population decimated by Te Rauparaha and very willing to welcome the new arrivals. Te Rauparaha and Te Rangihaeata were firmly opposed to Pakeha intrusion, and the latter chief was a leading figure in the skirmishes between Maori and European which occurred in 1845-6. Although this difficulty was soon overcome, other problems were equally serious and more persistent, especially the lack of suitable land and the thick forest which covered the district.

The Rimutaka Ranges between Wellington and the Wairarapa, prevented expansion on to the open land only a few miles to the north of the new settlement. In 1844 a difficult coastal route was pioneered and pastoralists spread over the region, but it was not until 1857 that a road was completed over the Rimutakas, and 1878 before New Zealand's only Fell railways linked the two regions. Settlement north of Masterton was hampered by the Forty and Seventy Mile bushes, and only in the 1870s did large-scale clearing begin. In the southern part of the region, small-farm settlements established in the 1850s led to a conflict between the pastoralists and the new men with limited capital.

Expansion in the Wellington district was slower, with dairying carried out closer to the city, and the steeper hill country used for sheep raising. The region's whole development has been shaped by Wellington's position as capital since 1865. As a result urban growth has been largely independent of the expansion of the rural hinterland.

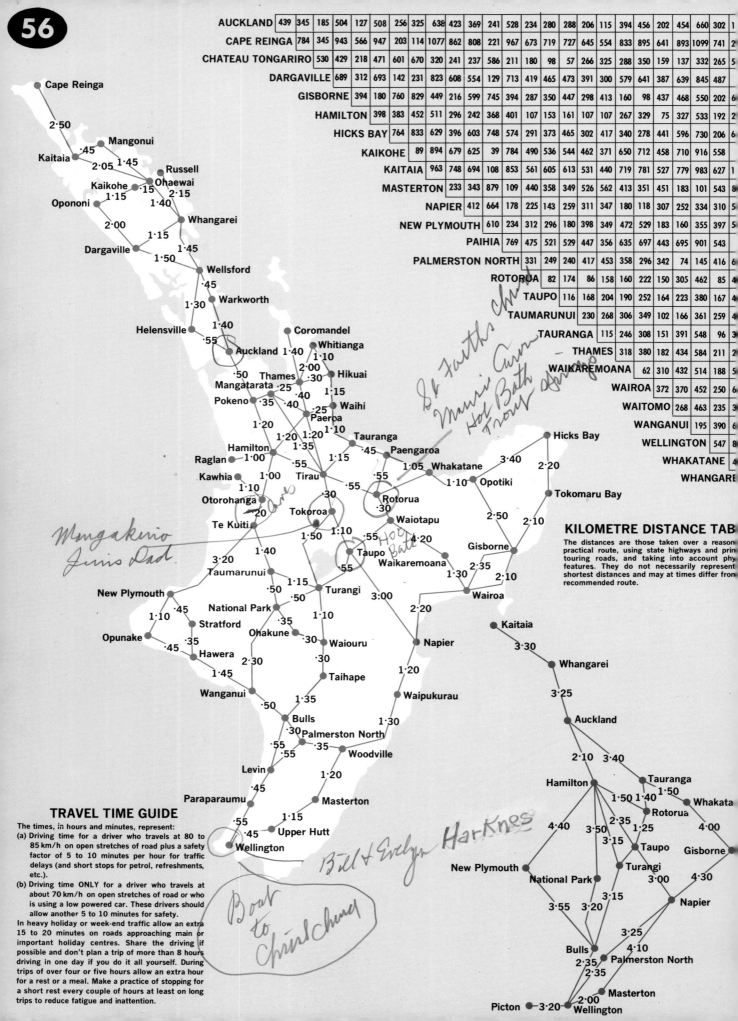

AUCKLAND	439	345	185	504	127	508	256	325	638	423	369	241	528	234	280	288	206	115	394	456	202	454	660	302
CAPE REINGA	784	345	943	566	947	203	114	1077	862	808	221	967	673	719	727	645	554	833	895	641	893	1099	741	
CHATEAU TONGARIRO	530	429	218	471	601	670	320	241	237	586	211	180	98	57	266	325	288	350	159	137	332	265		
DARGAVILLE	689	312	693	142	231	823	608	554	129	713	419	465	473	391	300	579	641	387	639	845	487			
GISBORNE	394	180	760	829	449	216	599	745	394	287	350	447	298	413	160	98	437	468	550	202				
HAMILTON	398	383	452	511	296	242	368	401	107	153	161	107	107	267	329	75	327	533	192					
HICKS BAY	764	833	629	396	603	748	574	291	373	465	302	417	340	278	441	596	730	206						
KAIKOHE	89	894	679	625	39	784	490	536	544	462	371	650	712	458	710	916	558							
KAITAIA	963	748	694	108	853	561	605	613	531	440	719	781	527	779	983	627								
MASTERTON	233	343	879	109	440	358	349	526	562	413	351	451	183	101	543									
NAPIER	412	664	178	225	143	259	311	347	180	118	307	252	334	310										
NEW PLYMOUTH	610	234	312	296	180	398	349	472	529	183	160	355	397											
PAIHIA	769	475	521	529	447	356	635	697	443	695	901	543												
PALMERSTON NORTH	331	249	240	417	453	358	296	342	74	145	416													
ROTORUA	82	174	86	158	160	222	150	305	462	85														
TAUPO	116	168	204	190	252	164	223	380	167															
TAUMARUNUI	230	268	306	349	102	166	361	259																
TAURANGA	115	246	308	151	391	548	96																	
THAMES	318	380	182	434	584	211																		
WAIKAREMOANA	62	310	432	514	188																			
WAIROA	372	370	452	250																				
WAITOMO	268	463	235																					
WANGANUI	195	390																						
WELLINGTON	547																							
WHAKATANE																								
WHANGARE																								

KILOMETRE DISTANCE TAB

The distances are those taken over a reason practical route, using state highways and prin touring roads, and taking into account phy features. They do not necessarily represent shortest distances and may at times differ from recommended route.

TRAVEL TIME GUIDE

The times, in hours and minutes, represent:
(a) Driving time for a driver who travels at 80 to 85 km/h on open stretches of road plus a safety factor of 5 to 10 minutes per hour for traffic delays (and short stops for petrol, refreshments, etc.).
(b) Driving time ONLY for a driver who travels at about 70 km/h on open stretches of road or who is using a low powered car. These drivers should allow another 5 to 10 minutes for safety.
In heavy holiday or week-end traffic allow an extra 15 to 20 minutes on roads approaching main or important holiday centres. Share the driving if possible and don't plan a trip of more than 8 hours driving in one day if you do it all yourself. During trips of over four or five hours allow an extra hour for a rest or a meal. Make a practice of stopping for a short rest every couple of hours at least on long trips to reduce fatigue and inattention.

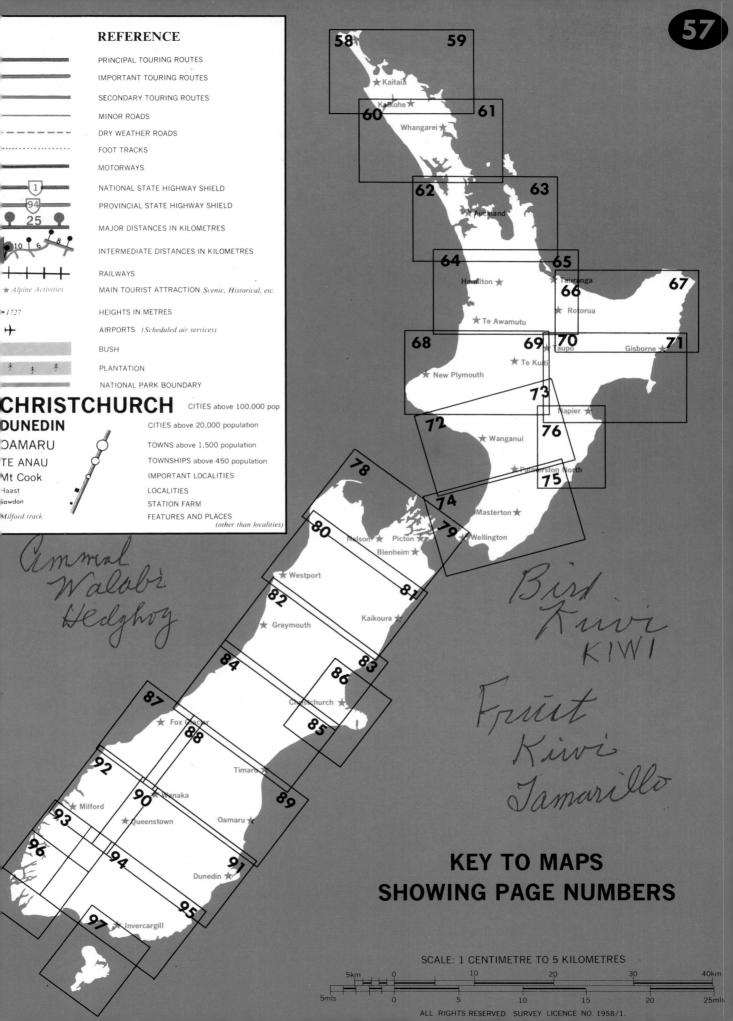

REFERENCE

PRINCIPAL TOURING ROUTES

IMPORTANT TOURING ROUTES

SECONDARY TOURING ROUTES

MINOR ROADS

DRY WEATHER ROADS

FOOT TRACKS

MOTORWAYS

NATIONAL STATE HIGHWAY SHIELD

PROVINCIAL STATE HIGHWAY SHIELD

MAJOR DISTANCES IN KILOMETRES

INTERMEDIATE DISTANCES IN KILOMETRES

RAILWAYS

MAIN TOURIST ATTRACTION *Scenic, Historical, etc.*

Alpine Activities

HEIGHTS IN METRES

AIRPORTS *(Scheduled air services)*

BUSH

PLANTATION

NATIONAL PARK BOUNDARY

CHRISTCHURCH
DUNEDIN
OAMARU
TE ANAU
Mt Cook
Haast
Sawdon
Milford track

CITIES above 100,000 pop

CITIES above 20,000 population

TOWNS above 1,500 population

TOWNSHIPS above 450 population

IMPORTANT LOCALITIES

LOCALITIES

STATION FARM

FEATURES AND PLACES
(other than localities)

KEY TO MAPS
SHOWING PAGE NUMBERS

SCALE: 1 CENTIMETRE TO 5 KILOMETRES

A　　　　　B　　　　　C　　　　　D

173°

1

Surville Cliffs

CAPE REINGA *Toponipotu Bay*　Hooper Pt.　*Tom Bowling Bay*　NORTH CAPE
Cape Reinga　*Spirits Bay*　Kapowairua
TE PAKI　108
Te Paki　Pandora
310
Motuopao Is
Cape Maria van Diemen　16
FARM PARK　Te Paki　5　11　Te Hapua
Twilight Beach　Waitiki Landing　*Waitiki*　Paua
Scott Point
Parengarenga Harbour
Thoms Landing
(*Karatia*)　*Ohao Point*

Motupia Is　25
19　AUPORI
Tangoake
L. Wahakari
2　Ngatiwhetu　Te Kao　*Great Exhibition Bay*
The Bluff
PENINSULA
24
Rarawa Beach
Ngataki　*Henderson Bay*
Grenville Point
Waihopo　Cape Karikari
Houhora　*Farmer Point*　*Moturoa Islands*　*Matai Bay*
Ninety Mile Beach　Pukenui　*Mt Camel*　*Rangaunu Bay*　*Whangatupere Bay*
47　Raio　335　*Karikari Bay*　6　*Knuckle Point*
Perpendicular Pt　Merita
Whatuwhiwhi
Rangiputa　Tokerau
3　Hukatere　Motutangi　*I. Waikaramu*　12　*Doubtless Bay*
32　Kaimaumau　Waingakau　*Takerau Beach*
L. Waiparera
Waiharara
Paparore　Lake Ohia　*L. Ohia*　4　10　M
Rangaunu Harbour
Pekerau　15　Parapara
35°　Te　Waimanoni　15　Kaingaroa　71
Waipapakauri Beach　Waipapakauri　California　Paranui　*Apuerew*
Sweetwater　(*Kareponia*)　Oruru
AWANUI　314　*Ohui R.*
1　Fairburn　Peria
Oturu　Rangitihi　*Puhi S.*　Kaiaka　38
4　10　KAITAIA　Pamapuria　Mangatoetoe
8　Victoria Valley　*Maungataniwl*
Ahipara Bay　Pukepoto　11　419　Te Rore　537　30　576
Reef Point　AHIPARA　Wainui Jct　*Taumatamahoe*　Diggers　Takahue　*Raetea*　381
Tauroa　3　976　Valley　473　750　*Mangatipa*
15　652
Manukau　512　Waiotehue　Mangamuka Bridge
448
5　Awaroa　21　Pukemiro　Broadwood　Orawau　10
Herekino　Puhata　12　11　*Tutekel*
Herekino Harbour　Maruroa　Papanga　Te Karae　*Kunik*
Owhata　5　489　*Uruugalo*
Whangape　11　Runaruna　22　Mata
350　Rotokakahi　731　Te Huahua　19
Pawarenga　722　18　*Tapuwae*　Kohukohu
Whangape Harbour　653　Panguru　Motuti　*Matawhera*　RAWI
496　Wairela　*Mouakuti*　*Opoke*
Mitimiti　21　Te Karaka　*Opara*
Reena

173°

A　　　　　B　　　　　C　　　　　D

1

2

3

Motukahakaha Bay
Whangaroa Bay
Stephenson Is
Flat Is
Cavalli Islands
Taupo Bay
Tauranga Bay
Te Ngaire
Motukawanui Is
Akatere
Totara North Saies
Wainui
Matangirau
Matauri Bay
Kahoe
Whangaroa
Te Huia
Te Pene
35
Waitaruke
Vaihapa
KAEO
Otoroa
angaroa
Pupuke
Ororua
Takou Bay
Takou Bay
Omaunu
Matawherohia
Upokorau
Otaha
Tavonui Bay
Te Ti'i
Puterua
Cape Wiwiki
CAPE BRETT
Waiare
Te Whau
Piercy Is
PUKETI
Pungaere
258
BAY OF ISLANDS
Urupukapuka Is
FOREST
Waipapa
Moturoa Is
Game Fishing
Deep Water Cove
Kerikeri Inlet
4
Puketi
KERIKERI
Kerikeri Inlet
Moturua Is
Rawhiti
Scen
Onewhero Bay
Parekura
Whangamumu
Waihou Valley
Puketona
Historical
RUSSELL
Bay
Waitangi
Haruru
Orongo Bay
Manawaora
Waimate North
PAIHIA
Paroa
Ngaiotonga
Okaihau
Te Ahuahu
Puketona
Opua
Bay
Ngaiotonga
Lake Omapere
Ohaeawai
Ferry
Tutaematai
Home Point
Utakura
L. Omapere
Whangae
Hupara
Waikare
Waihaha
Waikare
Whangaruru Beach
Whangaruru North
Remuera
Pakaraka
Waihaha
Whangaruru South
Bland Bay
Mangataraire
Settlement
MOEREWA
Taumarere
Punaruku
Cape Home
KAIKOHE
Ngawha
KAWAKAWA
Karetu
Te Rangi
Pukemoremore
Mokau
Mokau Bay
Ngawha Springs
Waiomio
Whangaruru South
Rakautao
Kawiti
Otiria
Caves
Helena Bay
Rimariki Is
Te Iringa
Tuhipa
Pokapu
Mimiwhangata
Ngepuhi
Punakitere
Ngapipito
Ruapekapeka
Tapuhi

E F G H

175°

1

A

Home Point
Tutaematai Waikare
Waihaha Whangaruru Beach
Karetu 28 312 Whangaruru North
31 Whangaruru South
Punaruku Cape Home
Te Rangi Pukemoremore Mokau Dakura Bay
408 390 Mokai Bay
Helena Bay Rimariki Is
Ruapekapeka Huruiki Mimiwhangata
Tapuhi 377 Puhipuhi 462

Akerama 20 Kaimamaku Otara Bay
Hukerenu Opuawhanga Whananaki
Waiotu 16 Whananaki South
Whakapara 9 Paetawai
onui Otonga Marua Waipaipai Matapouri
12 312 15 Tutukaka
Tanekaha 366 HIKURANGI Kaiatea 10
West 12 Matarau Kamo Springs 26 Ngunguru
Gumtown 11 Kiripaka
Kara Glenbervie Brynavon
kopu Whareora Tahere Horahora Ngunguru Bay
oroti 360 Mount Tiger Waiparera
Maunu WHANGAREI Pataua Taiharuru
APERE 12 Owhiwa Taraunui
Toetoe Waikaraka Hukuwai
Otuhi Otaika Valley Parua Bay
374 Otaika Tamaterau
Puwera 13 McLeods Bay
akaramea 13 Oakleigh PORTLAND Whangarei Heads
One Tree Point Taurikura
11 Takahiwai 490 Ocean Beach
Mangapai Marsden Bay BREAM HEAD
RANGE 421 Moewhare 15 MARSDEN POINT
Tauraroa Mata 12
Parahaka Springfield 271 Ruakaka The Chickens Is
15 274 56
Waiotira 1 Bream Bay Hen and Chickens Group
Waikiekie Ruarangi 10
18 335 The Hen Is
Taipuha 338 WAIPU Waipu R Sail Rock 36°
333 Braigh
81 Mareretu 257 Waipu Cove
Ararua 26 Langs Beach Bream Tail
10 292 Pilbrows Hill 396 179 Mangawhai Heads
Paparoa 12 Wairere 324 430 Molesworth Mangawhai Estuary
76 Brynderwyn 12 Cattle Mt
Huarau MAUNGATUROTO Tara
Matakohe Bickerstaffe Pukekaroro 20 Mangawhai
Pahi 1 Marohemu Hakaru
Whakapirau Ranganui 13 Te Arai Point
13 Kaiwaka
22 Arapaoa Te Arai
Batley Tanoa 13 Tomarata
Tinopai Topuni 7
Te Hana
KAIPARA Oruawharo Pakiri Goat Is
Onerin 10 Waiteitei 21 CAPE RODNEY
Port Albert 10 373 335 10
Wharehine Whangaripo Leigh
Hoteo Big Omaha Whangateau
Tapora North 15 Wayby 341 385 Ti Point
5 Cleaseby 23 Omaha Flats
Pouto Hill Waiwhiu The Dome Matakana Takatu
South Head 313 Dome Valley Tokatu Point
Tauhoa Hoteo Sandspit HAURAKI GULF
16 24 WARKWORTH
Mangakura Kaipara Flats Kourawhero Snells Beach 152
Kaipara Mullet Point Mansion House Bay
304 Mt Auckland Hills 306 Pohuehue Kawau Point
South Head Glorit Woodcocks 357 Martins Bay Flat Rock 175°

Poor Knights Islands

JELLICOE CHANNEL

Ngatamahine Pt
Little Barrier Island
Mt Hauturu 722
Te Titoki Pt East Cape

Craddock Channel

Horn Ro

Kawau Island

2

3

4

5

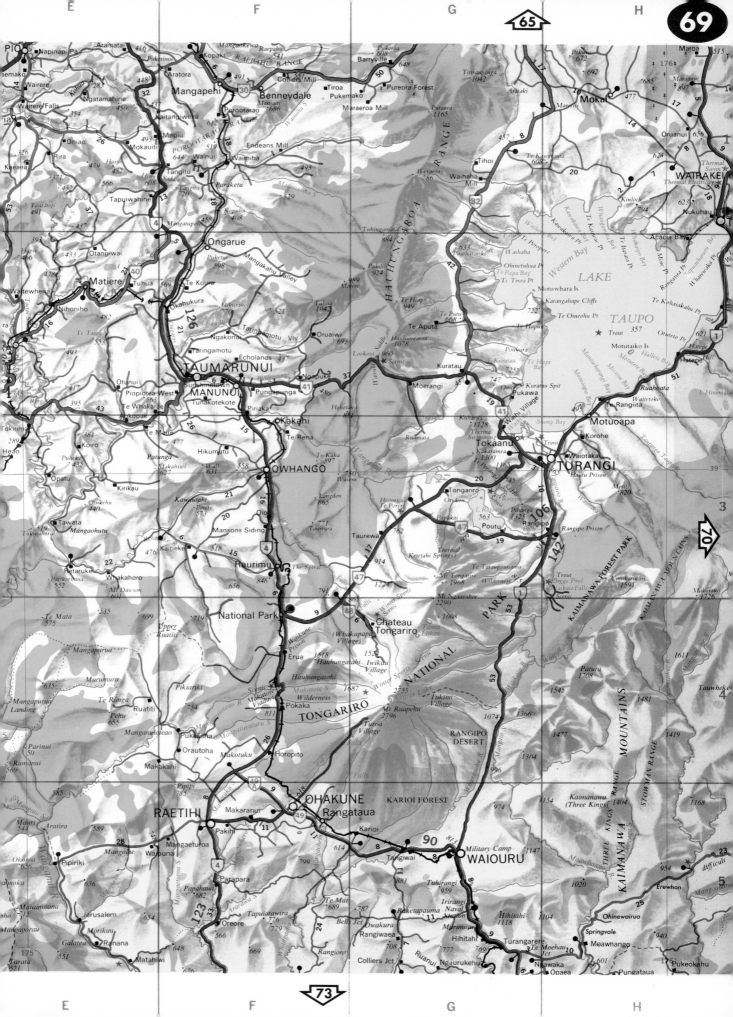

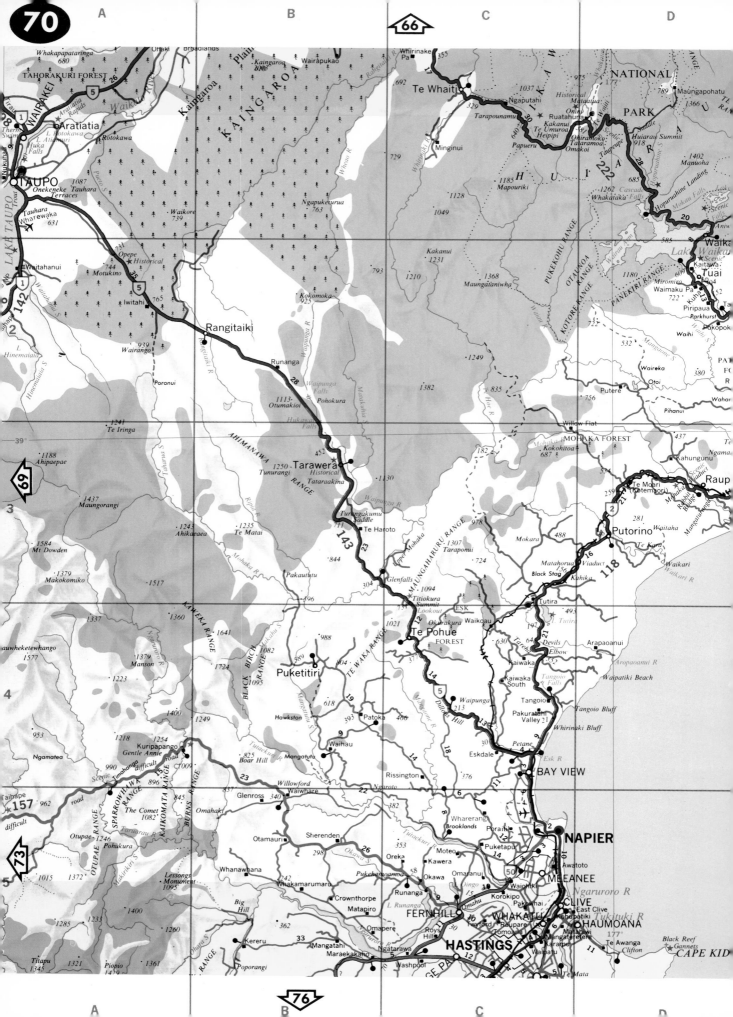

A   **B**   **C**   **D**

Otakeho

Kaupokonui   OKAIAWA

Fraser Rd

MANAIA

Ararata   366

Ihaha

Ketemarae   Moeroa

NORMANBY

Kohuratuahine   Tangahoe Valley

Tokaora

Kaupokonui R.

Kaupokonui Beach

Kapuni R.

Tawhiti   Mangaone   427

HAWERA

Taiporohenui   Meremere

Waingongoro R.

Ohawe Beach

1

Whareroa   Ohangai

Waihi Beach

SOUTH TARANAKI BIGHT

Tangahoe R.

Mokoia   Whakamaramara   411

Whakamaramara

Manutahi   Manutahi Sn   Hurleyville

Alton   Opaku

Hall Rd Sn   Karahaki

Kakaramea   Manukuroa

Otautu   Kohi   Onahina

Pariroa Pa

PATEA   Whenuakura   Okotuku   Mangawhio

Patea Beach   Whenuakura Pa

Patea R.   WAVERLEY

Whenuakura R.   Rangikura   Moumahaki

2

Ngutuwera   Rangitatau

Waverley Beach   Waitotara Sn   Waitotara

Nukumaru

Maxwell

Pukarakia Pa

Waitotara R.

Mowhanau   Rapar

3   **WANGAN**

Wanga

40°

174°

4

174°

5

175°

A   **B**   **C**   **D**

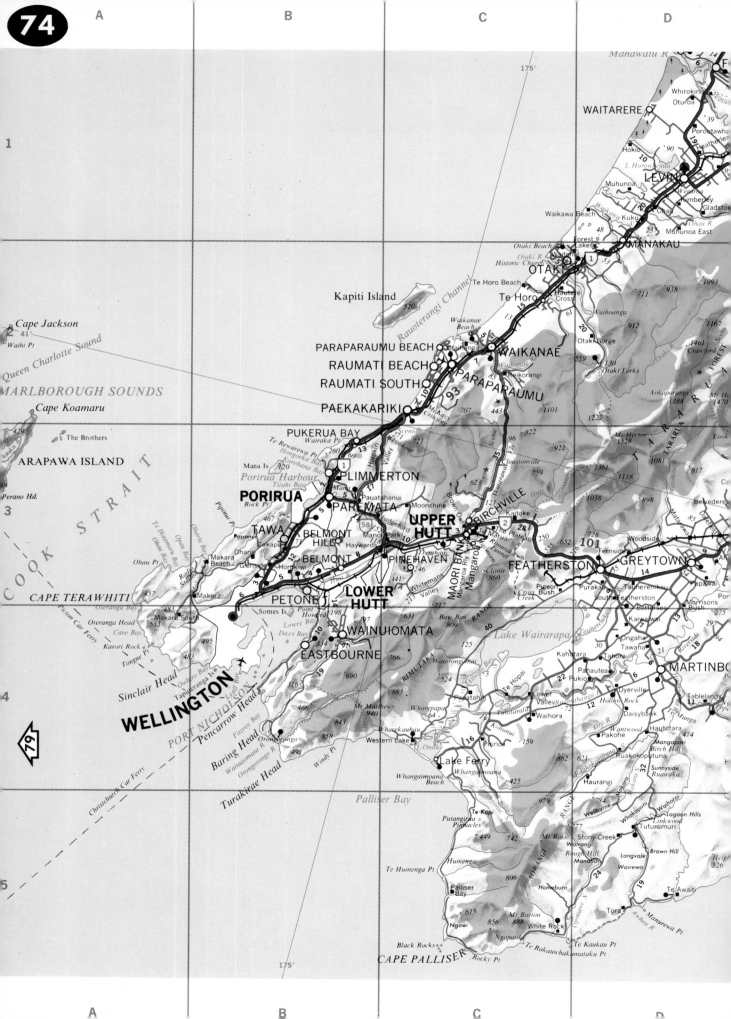

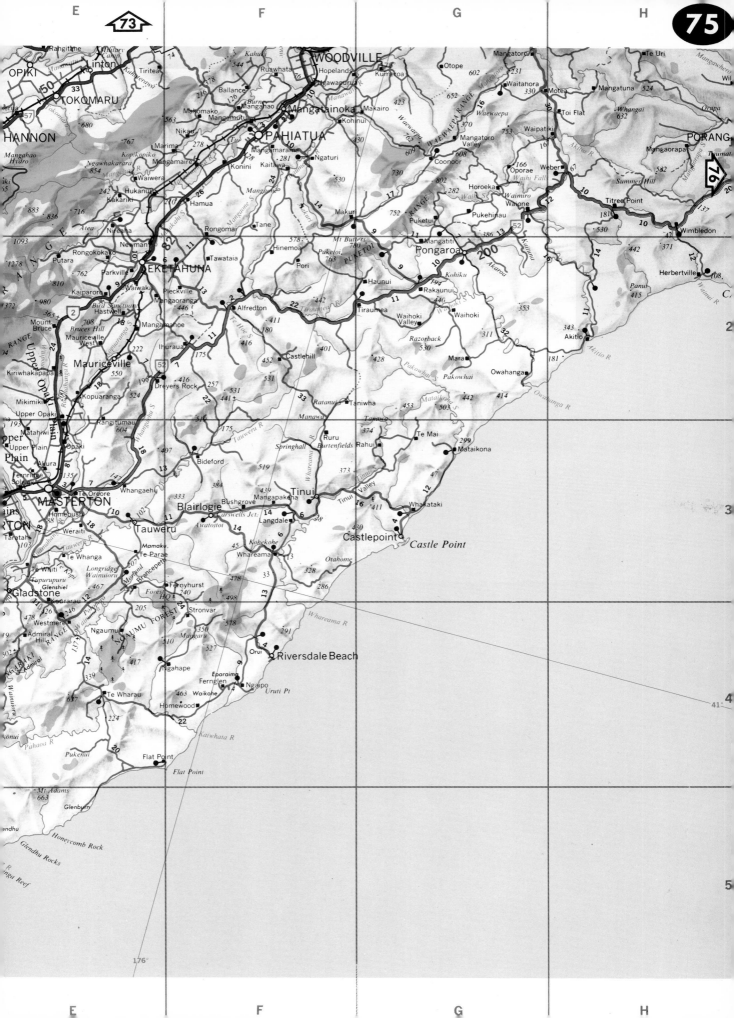

KILOMETRE DISTANCE TABLE

The distances are those taken over a reasonable, practical route, using state highways and principal touring roads, and taking into account physical features. They do not necessarily represent the shortest distances and may at times differ from the recommended route.

• indicates by Haast route

63	86	307	249	808	93	795	223	876	242	370	638	202	•512	590	231	•552	136	•373	190	964	455	232	767	544	ALEXANDRA
99	528	252	•801	320	•645	468	336	388	412	•922	311	668	102	263	383	98	602	241	451	476	150	698	440		ARTHUR'S PASS
50	736	475	964	102	798	28	559	116	643	1085	129	891	364	262	645	324	825	503	674	251	312	921			BLENHEIM
85	318	446	187	962	217	949	362	1037	474	308	792	30	•744	744	463	•784	96	•605	247	1118	609				BLUFF
33	424	163	652	353	486	340	247	428	331	773	183	579	288	135	738	248	513	427	362	509					CHRISTCHURCH
14	844	672	1117	206	•961	245	756	135	840	1238	380	1088	418	394	699	384	1022	582	871						COLLINGWOOD
95	276	199	290	715	283	702	115	790	331	411	545	217	702	497	•421	742	151	•563							DUNEDIN
80	287	493	•560	401	•404	579	596	469	498	769	508	•575	139	395	142	179	•509								FRANZ JOSEF
89	222	350	139	866	169	853	266	941	378	260	696	66	648	648	367	688									GORE
01	466	350	•739	222	583	352	434	290	510	•860	329	•754	40	216	321										GREYMOUTH
22	145	418	•418	543	262	673	376	611	356	539	650	433	281	537											HAAST
18	559	298	787	238	621	290	382	306	466	908	133	714	256												HANMER SPRINGS
41	426	354	699	262	543	392	438	330	514	•820	369	•714													HOKITIKA
55	278	416	157	932	187	919	332	1007	444	278	762														INVERCARGILL
31	607	346	835	231	669	157	430	245	514	956															KAIKOURA
61	394	610	121	1126	291	1113	526	•1150	550																MILFORD SOUND
64	211	211	484	684	328	671	216	759																	MT COOK
26	756	591	1029	118	•873	110	675																		NELSON
80	231	84	472	600	316	587																			OAMARU
8	764	503	992	130	826																				PICTON
84	117	335	170	839																					QUEENSTOWN
68	777	516	1005																						ST ARNAUD
40	273	489																							TE ANAU
97	273																								TIMARU
67																									WANAKA
																									WESTPORT

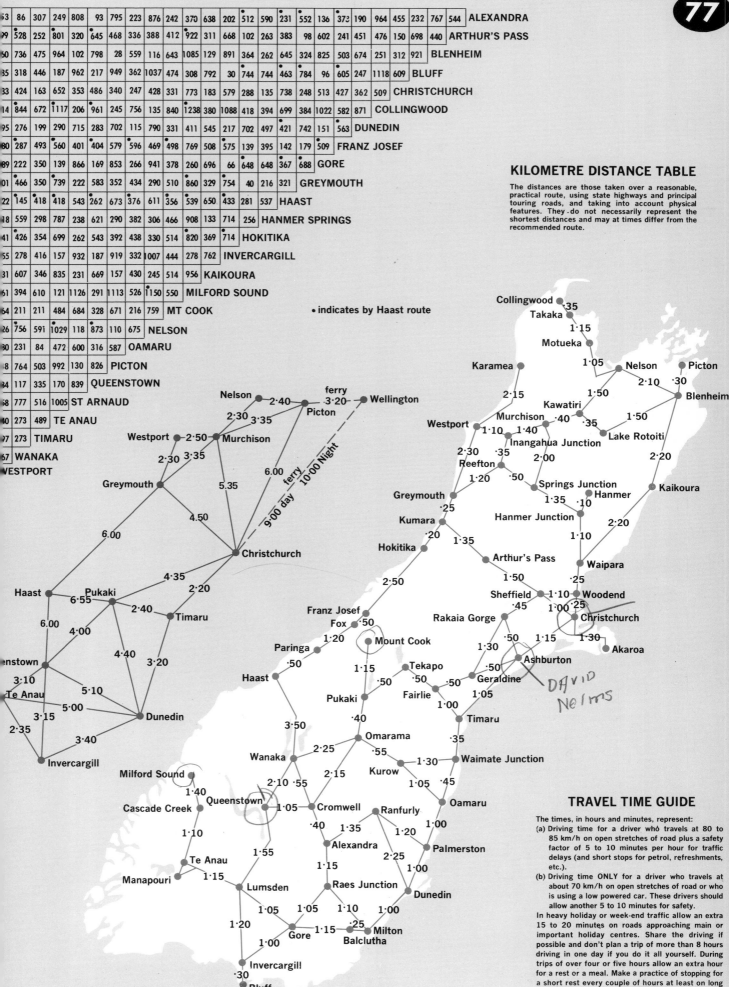

TRAVEL TIME GUIDE

The times, in hours and minutes, represent:

(a) Driving time for a driver who travels at 80 to 85 km/h on open stretches of road plus a safety factor of 5 to 10 minutes per hour for traffic delays (and short stops for petrol, refreshments, etc.).

(b) Driving time ONLY for a driver who travels at about 70 km/h on open stretches of road or who is using a low powered car. These drivers should allow another 5 to 10 minutes for safety.

In heavy holiday or week-end traffic allow an extra 15 to 20 minutes on roads approaching main or important holiday centres. Share the driving if possible and don't plan a trip of more than 8 hours driving in one day if you do it all yourself. During trips of over four or five hours allow an extra hour for a rest or a meal. Make a practice of stopping for a short rest every couple of hours at least on long trips to reduce fatigue and inattention.

A   B   C   D

1

173°

Lighthouse★ SPIT

FAREWELL

2

Pillar Pt.
Port Puponga
CAPE FAREWELL
Pilch Pt.
Puponga
Nguroa Bay
Mt Beale•288
Pakawau
Pakawau Inlet
Kaihoka

GOLDEN   BAY

Totaranui
Park HQ
Wainui
Bar Pt.
Ngatimiwha Inlet
South Head
Rakopi
Anatimo
14
Tonga Island
Tonga Roadstead
Sharks Head
Ferntown
Ruataniwha Inlet
Collingwood
Ligar Bay
Tarakohe
Abel Tasman
Memorial
ABEL TASMAN
Awaroa
Parapara Inlet
Pohara
26
Motupipi Inlet
Clifton
Mangarakau
Waitapu Inlet
Parapara
Onekaka
Rangihaieta
Rotorai
Torrent Bay
NATIONAL PARK
Aorere
Rockville
50
Patons Rock Waitapu
Motupipi
Paturau River
Puramahoi
27
Adele Island
The Blowhole
30
TAKAKA
Central Takaka
TASMAN   BAY
Otuhie
Waikoropupu
Pupu Springs
Anatoki
Kotinga
Caaman Road
Marahau
Bainham
Parapara Peak•1249
East Takaka
Sandy Bay
Kaiteriteri
15
Silverstream
Hamama
84
22
Stephens Bay
Paradise Peak•1537
Boulder Lake
Uruwhenua
Takaka Hill
35
Kairuru
RIWAKA
Centre
454
WAKAMARINA
Riwaka Valley
Umukuri
MOTUEKA
Gorton Downs•1113
DEVIL 1775 RANGE
Upper Takaka
Brooklyn
Port Motueka
Eureka•697
Mt Perry•1216
Devil River Peak
Mariri
Kina
HAUPIRI
Lower
Moutere
Tasman
37
50
NORTH   WEST   NELSON
DOUGLAS
Kakapo Peak
1769
1282•Hailes Knob
Rocky River
Pangatotara
Braeburn
Ruby Bay
Rabbit Is
30
Cobb River
Hoary Head•1462
Waiwhero
32
22
Harakeke
Mapua
Appleby
Tubman Hill•889
LOCKETT
L. Lockett
Ngatimoti
Rosedale
Mahana
Upper
Moutere
Redwoods
Valley
Heaphy
Track
Island L.
L. Sylvester
Graham Valley
Neudorf
Blackbird
Valley
Sunrise Valley
Eves Valley
Mt Ross•1372
Flanagan•1110
L. Henderson
Orinoco
Pokororo
Waimea West
FOREST PARK
Mt Cobb•1707
PEEL
RANGE
Woodstock
Dovedale
Win Valley
Pigeon Valley
Mt Domett•1615
TASMAN
58
Thorpe
Awa
WAKEFIELD
MOUNTAINS
ARTHUR RANGE
Stanley Brook
51
26
Belgrove
Foxhill
41°
Baton
Glen Rae
Tapawera
Pretty Bridge
Valley
Marawera
54
Clarke Valley
Waiti
Valley
Mt Olive•1427
Rakau
Kohatu
Mt Star•1539
Motupiko
24
Crow
Mt Jones
527 Mt Norris
Mt Garibaldi•1524
Matariki
Tadmor
GOLDEN DO

A   B   C   D

A B C D

1

Force River

Mt Oakden 1615

ROLLESTON RANGE

Mt Bryce 2309

Mt Meta 1023

Mt Algidus 1404 Algidus

Mt Misery 171

Mt McWhirter 2118

Mt Frieda 2048

Bond Pk

RIGGED RANGE

Double Hill

Glenfalloch

TOTARA STATE FOREST

Ross

Fraser Pk 1164

Mt Beauham 2144

Godley Pk 2087

Fergusons

MIKONUI STATE FOREST

Park Dome 2302

Totara Pk 1960

Smite Pk 1990

1972 Lagoon Pk

Kakapotahi

Mikonui R

Waitaha

Pukekura

Mt Hitchin

Mt Whitcombe 2644

2083

The Marquee

43°

IANTHE STATE FOREST

Mt Bonar

Scenic

Evans Creek

SMYTH RANGE

Mt Durward

Mt Smyth

Staces H 1478

Sugar L.

L. Heron

2

Wanganui Bluff

WANGANUI STATE FOREST

139

HILBERG RANGE

Ashburton R

Pyramid 1595

Lower Lake Heron

Cameron R

SALTWATER STATE FOREST

HARIHARI

Mt Tyndall 2524

BIG HILL RANGE

Clent Hills

Saltwater Lagoon

POERUA STATE FOREST

L. Rotokino

Mt Farrar 2432

Branch

Abut Head

Whataroa River

Rotokino

Te Taho

Mt Adams 2222

CLOUDY PEAK RANGE

Erewhon

Mt Potts

1334

3

OKARITO STATE FOREST

Whataroa

Gunns

Mt Price 1054

McClure Pk 2496

Havelock R

Okarito Lagoon

L. Wahapo

The Forks

63

Mt Loughnan 2576

Mesopotamia

Okarito

Ahna 2500

2541

3 Mile Lagoon

Scenic

L. Mapourika

Potters Creek

Mt Ross 2357

2048 1940 2114

170°

Waiho River

Matare

Scenic

PARK

Mt Sibbald 2798

Godley R

TWO THUMB

Ben

4

Franz Josef Glacier

Elie de Beaumont 3108

Mt Hope 2090

2004

NATIONAL ALPS

Thelma Pk 2057

Mt Erebus 2282

RANGE

Malte Brun 3176

Mt Tamaki 2442

Mt Radove 2412

Lilybank

RICHMOND

Mt Musgrave 2246

WAIKUKUPA

Douglas Pk Gl 3080

Fox Gl

Mt Johnson 2699

2187

Mt Gerald

Mt Misery 2293

Fox Glacier

Scenic

Mt Tasman 3497

HALL RANGE

s Pt

STATE FOREST

FOX RANGE

Craig Pk 1913

Balfour Glacier

2621 The Abbott

Mt Hazard 2223

Drake Pk

1972

RANGE

Beach

23

Mt Cook 3764

Ball Hut

The Nuns Veil 2735

GAMMCK RANGE

Richmond

Cook R

26

Mt Perouse 3078

MT COOK RANGE

COOK

Godley Peaks

KARANGARUA STATE FOREST

Fox River

Lytle Pk 2251

Mt Sefton 3157

1909 Mt Wakefield

LENDIG RANGE

Mt Tamaki 2346

1531

Glenmore

Mt Hay 1772

5

Karangarua

6

1933 Ryan Pk

Hooker R

Mt Blackburn 2388

2366

Mt Hay

Mt Edwards 1916

WESTLAND

Copland R

Alpine Memorial

Mount Cook

MOUNT

Lake Alexandrina

Lake Tekapo

142

Manakaiaua

Mt Peculiar 1912

Mt Burn 2738

Mt Cook Stn

1093

Mt John

44°

Jacobs River

Karangarua R

L. Rototekoiti

Mt McDonald 1996

170°

Mt Hopkins 2682

Holbrook

Bay

Sawtan

Stream

A B C D

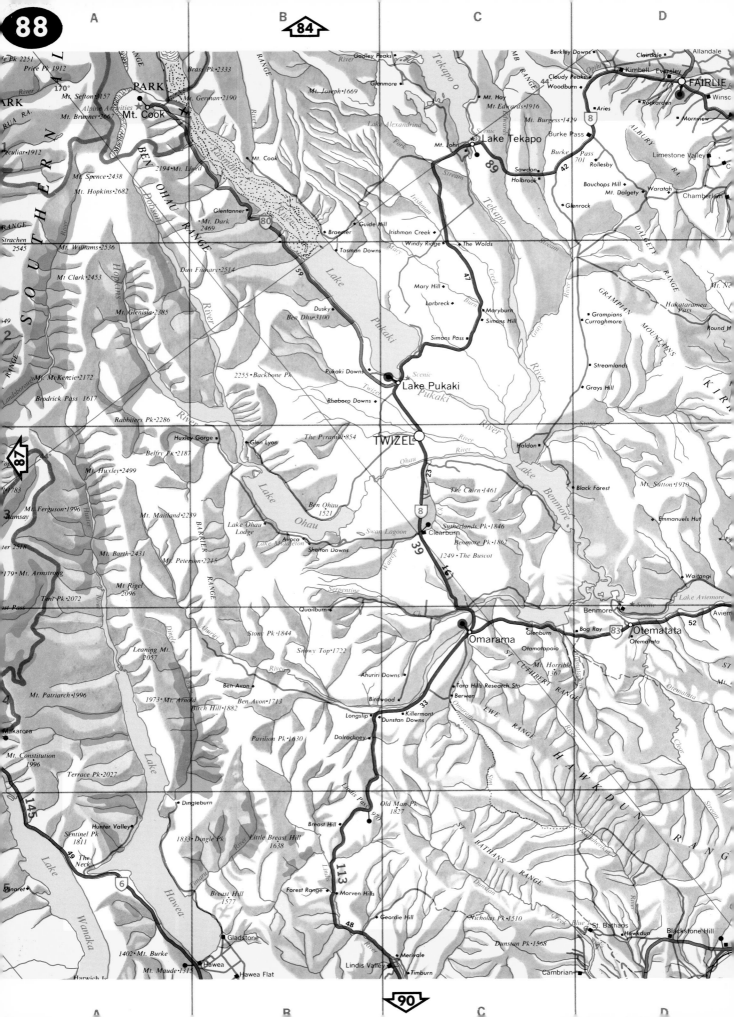

E   F   G   H

Upper Waitohi
Onuha
Opihi
PLEASANT POINT
171
Kerrytown
Seadown
Waitawa 19
Levels
Washdyke
8
Caroline Bay
TIMARU
Scarborough
Normanby
Salisbury
Kingsdown
Pareora
St. Andrews
Otaio River
Lyalldale
Esk Valley
Otaio
Springbank
Blue Cliffs
Kohika The Grange
Teschmaker
Makikihi River
Makikihi
Finlay Downs
Hunter
Bourndale
Watariari
Hook
Deep Creek Corner
Hook Bush
Gunns Bush
Studholme
Landsdown
Deep Ck.
Norton Reserve
Waitana
Mikuroa
Kelseys Bush
WAIMATE
Willowbridge
Uretane
Morven
Arno
Gum Tree Flat
Kapua
Green Hills
McLeans Broad Gully
Waihao Forks
Grays Corner
Waihao Downs
Dog Kennel Creek
Mt. Harris
Waikakahi
Glenavy
Salmon
Tawai
Elephant Hill
82
Ikawai
Clarksfield
Hilderthorpe
Grassy Hills
Uxbridge
Gibsons
Wharekuri
Hakataramea
Black Point
Bortons
Aitchisons
Papakaio
Richmond
Corrie
Mt. Parker
Georgetown
35
Awamoko
Peebles
Pukeuri Junction
Vaitaki
UROW
Hilles
Station Peak
Waitaki
Kokoamo
Airedale
Strachans
Otiake
Waikaura
23
Rosebery
Duntroon
Windsor Park
Ardgowan
Ngapara
Queens Flat
Corriedale
Otekaieke
Macrewhenua
Island Cliff
Tilverstowe
Windsor
Livestone
Enfield
Whitstone
OAMARU
Tokarahi
Post Office Gully
Weston
Cape Wanbrow
Clifton Falls
Alma
White Rocks
Tapui
Maruakoa
Five Forks
Kia Ora
Beatties Hill
Livingstone
Totara
Teschemakers
Fuchsia Creek
Kauru Hill
Incholme
Kakanui
Kakanui South
Maheno
Mt. Domet 1935
Ben Lomond 1052
Island Stream
All Day Bay
Danseys Pass
Kuriheka
Clareview
Mt. Kyeburn 1636
79
Waianakarua River
Herbert
Kyeburn Diggings
Glencoe
Waianakarua
Mt. Pisgah 1643
Siberia Hill 1274
59
Mt. Miserable 886
Hampden
Moeraki Boulders Moeraki Pt
MOUNTAINS
Moeraki
Hillgrove
25
Katiki
Longlands
26
Trotters Gorge
Naseby
Kyeburn
Swinburn Pk
Islay Downs Morrisons
The Brothers
HORSE
Razorback 603
Shag Point
164
Waihemo
21
85
Waynes
Shag Pt
Makareao
Bushey

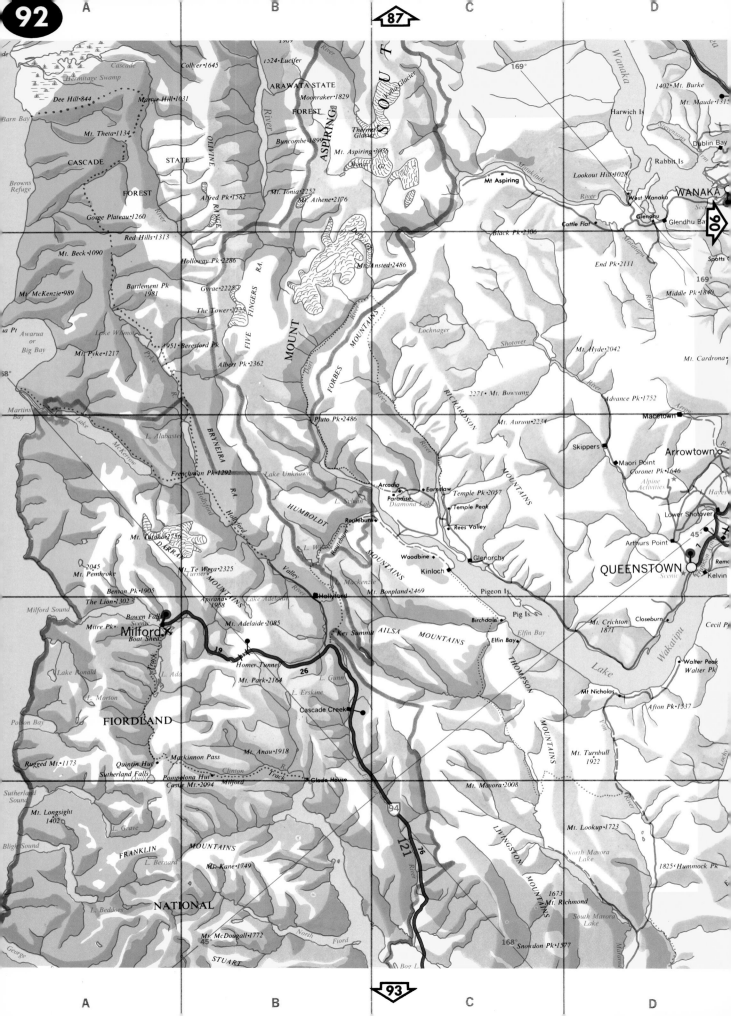

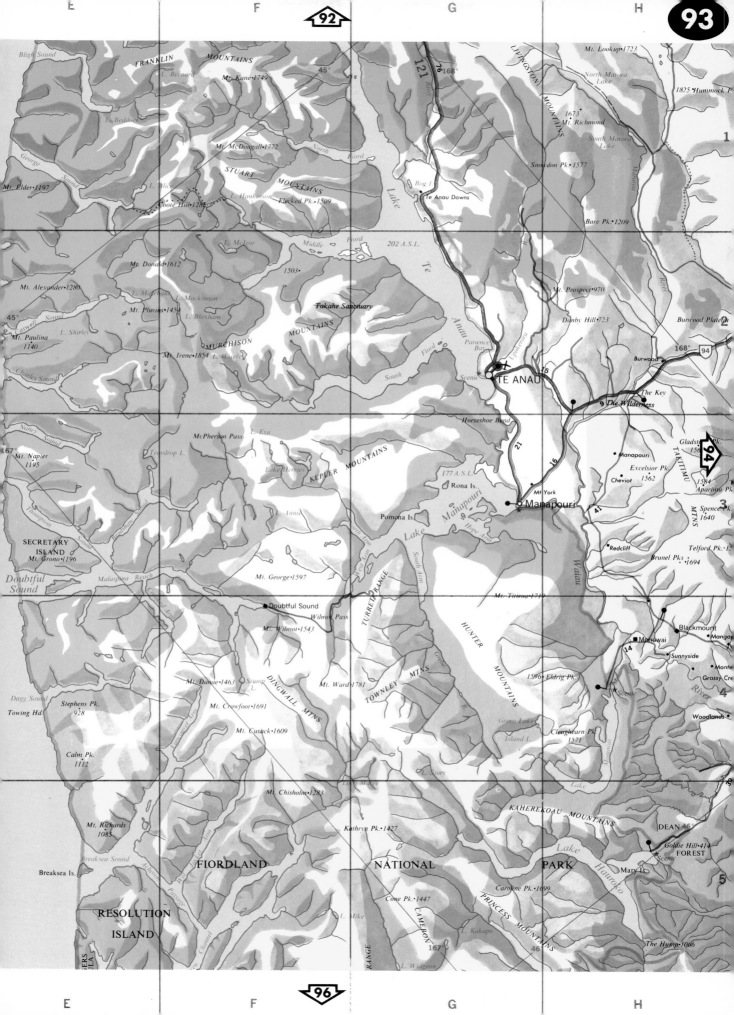

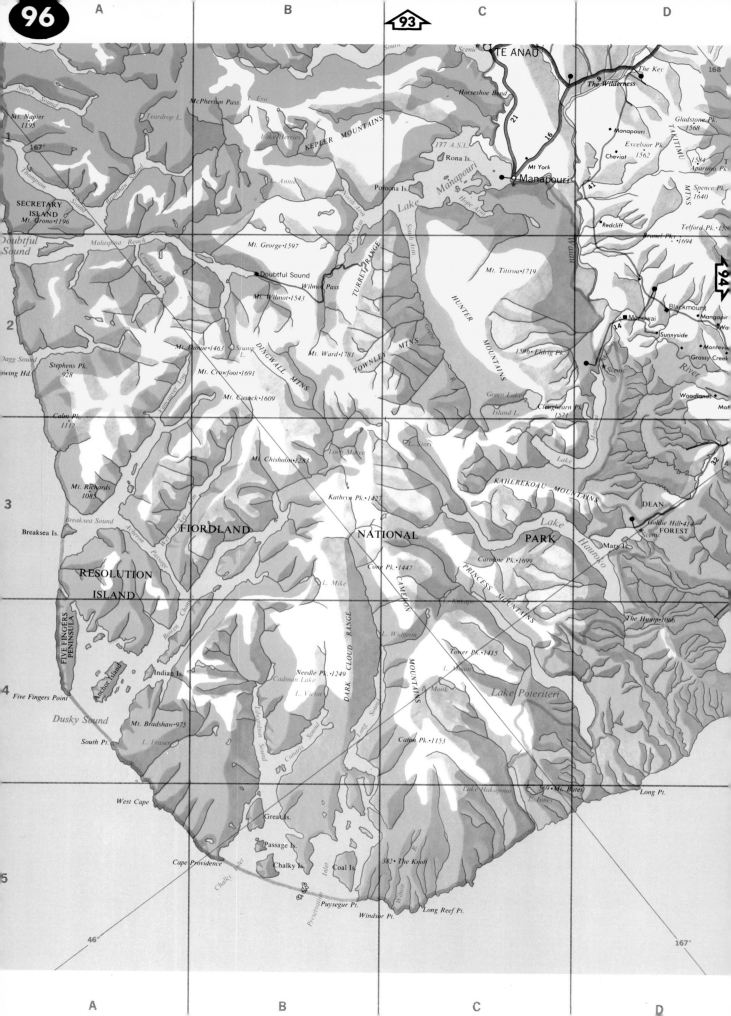

A B   C D

1

Nancy Sound
Mt. Napier
1195
167°
Thompson Sound
Bligh Sound
SECRETARY
ISLAND
Mt. Grono •1196

Teardrop L.
McPherson Pass
L. Eva
Lake Herries
KEPLER MOUNTAINS
L. Annie
North Arm

Horseshoe Bend
TE ANAU
The Key
The Wilderness
9
168
21
16
Manapouri
Gladstone Pk.
1568
Excelsior Pk.
1562
Cheviot
TAKITIMU
1584
Aparima Pk.

Doubtful Sound
Malaspina Reach
Crooked Arm
Mt. George •1597
Doubtful Sound
Wilmot Pass
Mt. Wilmot •1543
TURRET RANGE
Head Arm
South Arm
177 A.S.L.
Rona Is.
Pomona Is.
Lake Manapouri
Hope Arm
Mt York
Manapouri
Mt. Titiroa •1719
HUNTER
MOUNTAINS
41
Redcliff
Brunel Pk. •1694
Spence Pk.
1640
Telford Pk. •138
94

2
Dagg Sound
Wing Hd.
Stephens Pk.
928
Mt. Danae •1463
Stump L.
DINGWALL MTNS
Mt. Crowfoot •1691
Mt. Cusack •1609
Vancouver Arm
Mt. Ward •1781
TOWNLEY MTNS
Greene R.
1596 • Eldrig Pk.
Green Lake
Island L.
Clearhearn Pk.
1571
Monowai
14
Sunnyside
Scenic
Mangapir
Wa
Montevu
Grassy Creek
River
Woodlands
Mot

Calm Pk.
1112
Mt. Richards
1085
Mt. Chisholm •1283
Loch McKee
L. Story
KAHEREKOAU MOUNTAINS
Lake
Manowai
32
4

3
Breaksea Sound
Breaksea Is.
Acheron Passage
Wet Jacket Arm
FIORDLAND
Kathryn Pk. •1427
NATIONAL
Cone Pk. •1447
PARK
Caroline Pk. •1699
PRINCESS MOUNTAINS
Lake
Hauroko
DEAN
Goldie Hill •414
FOREST
Mary Is.
Scenic

4
RESOLUTION
ISLAND
FIVE FINGERS PENINSULA
Anchor Island
Indian Is.
Five Fingers Point
Dusky Sound
Mt. Bradshaw •975
South Pt.
L. Fraser
Bowen Chan.
Cadman Lake
Needle Pk. •1249
L. Victor
L. Mike
DARK CLOUD RANGE
Edwardson Sound
Cunaris Sound
CAMERON
MOUNTAINS
L. Widgeon
Tower Pk. •1415
L. Miquai
Mt. Monk
Long Sound
Caton Pk. •1153
The Hump •1066
Lake Poteriteri

5
West Cape
Great Is.
Passage Is.
Cape Providence
Chalky Is.
Coal Is.
Chalky Inlet
Preservation Inlet
Puysegur Pt.
Windsor Pt.
Wilson R.
Long Reef Pt.
382 • The Knoll
Lake Hakapoua
L. Innes
569 • Mt. Bates
Long Pt.
46°
167°

A B C D

Glenburn
Fairfax
168
Gropers
Gropers Bush
Ermedale
Thornbury
Gummies Bush
Opuaka
Waipango
Wild Bush
Longwood
Tihaka
Round Hill
Wakaputa
Kawakaputa Bay

Wrights Bush
Flints Bush
Hazletts
99
Waimatuku
Taramoa
Argyle Corner
West Plains
Waikiwi
92
Waimatua

INVERCARGILL
OTATARA
New River Ferry
New River
Estuary
Moturimu
Awarua
30
1
Greenhills
Ocean Beach
Omaui Beach
Oreti Beach
Bluff Harbour
265
BLUFF
Tiwai Pt
Tiwai Pt
Stirling Pt
Dog Is.

Riverton Rocks
Howells Pt.
RIVERTON
Colac Bay
Colac Bay
Oraka Pt.

Awarua
Bay
Waituna Lagoon
Waipapa Pt.

TOETOES BAY

Green Is.
Ruapuke Island
Bird Is.

FOVEAUX

Wakaputa Pt
Centre Is.

STRAIT

47°

Mt. Anglem •979

Port
William
Horseshoe Bay
Halfmoon Bay
Bench Is.

Halfmoon Bay
(Oban)
Scenic
Ulva Is.
The Neck

Codfish Is.

Paterson
Inlet
Big Glory Bay

Mt. Rakeahua
675

Adventure Hill •258
Port Adventure
Shelter Point
Breaksea Islands

STEWART ISLAND

Mason Bay

Table Hill •715

Doughboy Bay

Lees Knob •586

North Arm

Pearl Is.
Port Pegasus
Anchorage Is.
Noble Is.

Big Moggy Is.

Smiths Lookout
•35
Broad Bay

South Cape

Big South Cape Is.

Flour Cask Bay
South West Cape

47°

168°

1

2

3

4

5

SOUTH ISLAND

NELSON

Over 70 per cent of the Nelson region is classified as hilly or mountainous, only 16 per cent as cultivable. To the east are the Bryant Ranges which extend into the Richmond Range and the St Arnaud Range further south. Running down the centre of the region is the bulk of the Tasman Mountains and the Arthur Range. South of Murchison this mountainous land broadens, to block access to the West Coast and North Canterbury except through one or two routes. Springs Junction and Maruia Springs near Lewis Pass, are on the southern boundary. The west coast from Karamea to the sandy hook of Farewell Spit is largely inaccessible. Nearly all the population and economic growth is concentrated in the area of flat or reasonably rolling land which lies south of Tasman Bay.

History

The first European to venture into the Nelson region was Abel Tasman. In December 1642 he sailed into Golden Bay in search of water and food; instead he lost four crewmen to hostile Maoris and fled, naming the bay Murderer's Bay as a result.

For almost 200 years after that brief episode, the region was largely ignored by Europeans. Then in 1842 settlers sent out by the New Zealand Company arrived. The choice of Tasman Bay was not the Company's original preference. However Governor Hobson vetoed the proposed settlement of the Canterbury Plain and Nelson was selected for its harbour of sorts and the small extent of the Waimea Plains. Partly because of this compromise the first years of settlement were very difficult and starvation threatened on several occasions.

Because of the poverty of most of the settlers, and the scarcity of suitable land, a pattern of small, intensively farmed holdings developed.

In 1857 gold was discovered at Collingwood in Golden Bay and in seven years over $500,000 was won. Collingwood grew into a town of over 4,000 and at one stage was considered as the site of New Zealand's seat of government. This wave of euphoria soon passed. Instead Nelson settled down to a slower rate of growth based on the diversity of crops produced on the valley flats running back from Tasman and Golden bays.

Industry

Tourism has also become an important industry although it is not promoted with quite the same fervour as in other areas. With its pleasant, extremely sunny climate, the beaches of Golden and Tasman bays, and the more rugged charms of the Abel Tasman National Park and the Nelson Lakes National Park which includes the Heaphy Track, Nelson has many attractions for those seeking quiet relaxation. One of the more unusual attractions is a trip out to Cape Farewell and Farewell Spit, a long sandy arm protecting Golden Bay from the ravages of the Tasman Sea.

MARLBOROUGH

Three distinct area make up the Marlborough region. To the north is the drowned valley system of the Marlborough Sounds; south of this rugged, water-fretted area is the broad, almost flat area of the Wairau Plains which reaches out to the sea to the north-east. The Bryant Range separates Marlborough from Nelson to the west. This runs from north-east to south-west to link with the hill country which makes up the third, most southerly, part of the region. The back country ranges reach their greatest elevation in the Inland Kaikouras 2885 m and the Seaward Kaikouras 2609 m. South of Ward on the east coast, the Seaward Kaikouras drop abruptly into the sea.

History

Marlborough was a favoured spot for Maori settlement, partly because of its climate and fertile soil, and also because it lay on the route to the treasured greenstone on the West Coast. The region was occupied by a succession of conquering tribes, last, but not least, being the Ngati Toa under Te Rauparaha.

By then Marlborough had become the site of the first European settlement in the South Island. In 1827 Jack Guard set up a whaling station in Tory Channel, barely half a century after James Cook had used Ship's Cove in Queen Charlotte's Sound as a base during his three visits to New Zealand. Whaling from the Sounds persisted until 1865 when the last station in the country closed down.

The interior was largely unexplored until the 1840s when the expanse of the Wairau Plains was coveted by the land-hungry Nelson settlers. A misunderstanding between Captain Arthur Wakefield and Te Rauparaha in 1843, led to the brief but bloody clash now known as the Wairau Affray — the only Maori-European fighting in the South Island. This did not slow European expansion however. By the late 1840s the runholders had become established. In 1864 the Wakamarina Valley was the site of a brief goldrush, but sawmilling had a longer, though equally destructive history. By the turn of the century, Marlborough had settled into a pattern of pastoral farming based on sheep and dairying on the Plains and lower hills, with sheepfarming dominant in the back country.

Industry

Pastoral farming in the Marlborough Sounds has declined in recent years, although large sheep runs still persist in some areas. A more recent development has been the proliferation of holiday homes in many of the bays in the Sounds. Picton at the head of Queen Charlotte's Sound has grown as a holiday centre and as the southern terminal of the Cook Strait ferries, one of the most important links between the two main islands.

THE WEST COAST

The West Coast is a long thin strip of land bounded on the west by the Tasman Sea and to the east by the Southern Alps. Over 560 km long, it is less than 48 km wide throughout except to the east of Reefton where the region follows the Buller inland for about 80km. More than three-quarters of the West Coast is uninhabited forest and mountain.

In the southern part of the region are the Fox and Franz Josef Glaciers which until recently descended to within 213m of sea level. They are readily accessible by road.

History

The Maoris valued the West Coast for the greenstone it held, but even the prized nephrite could not compensate for the inaccessibility, high rainfall and rugged, bush-covered country, and the population remained small. Tasman made his first landfall off Cape Foulwind in 1642, and sailed north seeking more obviously hospitable shores. Later Europeans were equally reluctant to explore the area, and it was not until 1846 that Thomas Brunner, Charles Heaphy and Ekelu made one of New Zealand's epic explorations.

Development came to the coast with a rush in the mid-1860s. The discovery of gold led to an era which has passed into New Zealand history and folklore. The population grew in three years to about 30,000 and towns sprang up — and died overnight.

The peak of the gold days had passed by the 1870s, although gold mining remained important until the turn of the century. Instead coal and timber became king. By 1910 the West Coast was producing over 1.3 million tonnes of coal, more than 60 per cent of the country's total. Fifty years later output had declined to under 1 million tonnes, about one third of the national output, and has continued to drop. Although the energy crisis of the mid-1970s has revitalised the coal industry to a certain extent, it is unlikely that it will ever regain its former dominance.

Timber milling has also declined since the 1950s, although a hotly-disputed plan to mill the beech forest could make timber milling a major industry again.

Industry

Although one dredge is still working on the Taramakau River, gold makes a negligible contribution to the Coast's economy. Coal too will probably never return to its former importance, partly because of the cost of underground mining but also because of the need to conserve resources for future needs or emergencies. The only other known mineral deposit suitable for commercial exploration is the ilmenite ironsand found on the beaches north of Greymouth. Over 10 million tonnes could be recovered, but markets overseas are not favourable.

The plan to mill the beech forests mainly to the north of Hokitika could make it the largest single industry on the Coast. Part of the forest would be milled selectively to allow regeneration and part cleared for exotic forest planting. The controversy this scheme has caused could lead to its abandonment.

SOUTHLAND-FIORDLAND

The southern-most part of New Zealand has never experienced the sudden upsweeps and reversals of fortune undergone by other regions. Its history has been one of a long and successful campaign to tame the land. Today Southland is one of the most productive agricultural areas in New Zealand.

To the west the Hunter Mountains separate Southland from Fiordland, and mountain ranges also mark the region's northern limits. From the Garvie and Eyre Mountains, the southern-most extensions of the Southern Alps, the land falls away towards the sea, interrupted only by the Takitimu Mountains and the Hokonui Hills. In the north-western corner of the regions are two of New Zealands most beautiful and controversial lakes, Manapouri and Te Anau.

The eastern boundary roughly parallels the Mataura River a few km to the west. Within these margins are a number of alluvial plains, the heart of Southland's agriculture. Between the high country and the Hokonui Hills are the Five Rivers and Waimea Plains, drained by the upper Oreti and Mataura Rivers. To the west of these is the Southland Plain, bounded by the Mataura and Aparima rivers. The western-most area of flat land is the lower Waiau plain, separated from the Southland Plain by the Longwood Range.

In contrast, the Fiordland region contains some of the most inaccessible and sparsely populated country in New Zealand. The only road across is by way of the Homer Tunnel to Milford Sound, virtually the only centre of population in the area. Flat land is minimal, and some of the slopes are so precipitous that even plants find it almost impossible to grow. Despite this and the phenomenal rainfall, the long reaches of water contain some of the most beautiful scenery in New Zealand.

History

The first Polynesian settlers came to this area in search of the moa, which may in fact have survived in some remote pockets into the 19th century. Maoris too settled in the area, perhaps lured by the delicacies of the muttonbird and oyster. The impact of European discovery was more brutal. The first Pakehas to arrive in any quantity were sealers who clubbed the seals almost to extinction in a few brief years at the beginning of the 19th century.

More permanent settlements were established by the whalers who followed, many of them remaining to become the district's first farmers with extensive sheep farms. In 1856 there was a further influx,

100

largely from the eight year old settlement in Otago. Perhaps because of its parent-child relationship, Southland soon fell out with Otago and became a separate province in 1861, just in time to miss out on the wealth from the gold rushes which made Otago rich. But its parentage still survives in the region; it is perhaps more Scottish than Otago today.

Certainly the settlers brought with them a determination to succeed; the land has been made into one of the most productive parts of New Zealand despite the short summer and problems with soil fertility.

Fiordland was largely ignored by moa hunters and Maoris alike, only the remains of defeated tribes retreated into its vastnesses. James Cook made contact with a few of these when he anchored in Dusky Sound between March 26 and May 11 1770. By the time the first European settlers arrived, the area was uninhabited. Most famous of the Fiordland settlers was Donald Sutherland who came prospecting in the early 1870s. In 1887 his isolation was ended when Quintin Mackinnon discovered the Mackinnon Pass which gave land access to Milford Sound. Most tourists came by ship until the early 1950s when the Homer tunnel was opened. Much of this area is now part of the Fiordland National Park.

Industry

Farming is the basis of Southland's prosperity. Most important agricultural areas are the three plains' areas, in the lower Waiau, Southland and the Five Rivers-Mataura. Here dairying, fat lambs and sheep, and cattle are raised, with farmers showing considerable readiness to switch from one to the other according to economic conditions. Up to the mid-1970s the region showed a marked increase in beef, mutton, and lamb production, but wool output has been virtually static since the mid-1960s.

The waters to the south of the region has also been a source of prosperity. The total value of fish and shellfish landed is greater than for any other part of New Zealand. A large proportion of this value comes from the oyster beds in Fouveaux Strait, though reduced limits has cast some gloom over the industry.

CANTERBURY

Canterbury contains New Zealand's largest area of flat land and also its highest mountain peaks. From the diversity of landscapes has come an equally great variety of agricultural activities and industries. Its northern boundary is formed by the Waiau River; its southern limits by the Waitaki River. In contrast the eastern and western boundaries are marked by opposites: the Pacific Ocean to the east and upthrust of the Southern Alps to the west.

One third of the region is classed as flat most of it contained in the broad expanse of the Canterbury Plains. To the north of the Waipara River the area of the plain narrows with the encroachment of the foothills of the Alps, but to the south it extends almost undiminished to the Rangitata River. Behind the Canterbury Plain, the land rises into the foothills and then the ranges of the Southern Alps. Almost due west of Timaru, the hill country flattens into the plateau of the Mackenzie Country, with Mount Cook, New Zealand's highest mountain at 3764m, on the northern margin on the plateau area.

History

Moa hunters were the first settlers in the region, probably burning off much of the forest cover on the Plains. Compared to the North Island, the region was sparsely populated in classical Maori times, and by the 1850s when Europeans began to arrive in large numbers the local Maoris had been reduced to about 500 by disease and the ravages of Te Rauparaha.

One of the more colourful myths about New Zealand arose with the arrival of 80 French settlers at Akaroa on Banks Peninsula in August 1840. From this has come the story of a "race" between France and Britain to claim New Zealand, a story which has no foundation in fact. Today there are few remnants of the little pocket of France, apart from a few French names and some distinctive colonial architecture; but it was perhaps fitting that this little group, so alien from the strongly British settlers who were to claim the rest of Canterbury, should have settled Banks Peninsula which has a climate and topography completely different from the rest of the region.

In December 1850, 780 immigrants arrived from England under the auspices of the Canterbury Association to found a replica of an ideal rural society. The Association's dreams were realised in a way they had not anticipated. In the 1850s squatters and runholders spread rapidly across the plains and lowland hills, then after a brief pause, into the high country. Here they reproduced the life style of landed gentry in the midst of a raw colonial society and, despite the changes which have occurred since, their values and attitudes — and way of life — have persisted to colour the whole atmosphere of the region.

Canterbury never enjoyed the sudden stimulus of mineral discoveries experienced by Otago, though the markets for foodstuffs provided there, and to a lesser extent on the West Coast, by the gold rushes, did help growth in the 1860s. This decade was also marked by the spread of small farms, radiating out from Christchurch and Timaru further south.

The 1870s changed the nature of agriculture in Canterbury as double furrow steel ploughs, steam threshing machines and irrigation allowed the expansion of cultivation. Wheat became an important crop and helped cushion the region from the depression that afflicted the colony in the late 1870s and 1880s. The development of frozen meat in the 1880s gave a further stimulus to the spread of more intensive methods of farming, although on the high country, large runs of 16,000 to 61,000 hectares remained the pattern.

Industry

Farming has always been the basis of Canterbury's economy, with the Plain accounting for much of the area's output. Cropping is carried out from Culverden in the north to south of Timaru. On some farms the cropping is only meant to provide feed for fat lamb production, but others rely on cash cropping for their income. Mid and south Canterbury produces over half New Zealand's wheat crop. Dairying is usually combined with cropping and is largely concentrated around Christchurch, as is most of the market gardening, orcharding and small fruit output, most notably with lucerne production and cattle raising.

The high country, especially the Mackenzie Country is the preserve of the sheep runs, but their former isolation has been disturbed by recent developments. The search for sources of hydro electric power has invaded the Mackenzie Country and one of the more ambitious schemes ever undertaken will channel water from Lakes Tekapo and Pukaki to Ohau and thence down the Waitaki to augment new Zealand's power output. The runholders are also contributing to the invasion of their territory. Many have found that tourist accommodation can be as lucrative as the more traditional sheep farming.

OTAGO

Otago is the only region in the South Island to have boundaries on the east and west coasts. From a broad base on the Pacific side it drives across country in a narrowing wedge to the Tasman Sea, enclosing within its boundaries some of the most varied scenery in New Zealand.

The eastern boundary follows the coast from the Waitaki River south to the bottom "corner" of the South Island almost to the Mataura River. The southern boundary follows the Mataura River northwards, turns west to skirt the Garvie Mountains and the southern shore of Lake Wakatipu before threading through the mountainous country south of the Darran Mountains to the northern side of Milford Sound. The northern boundary begins at Awarua Bay, strikes eastwards to the base of Mount Aspiring then turns north to the Haast Pass. It then crosses the Southern Alps and runs south to the Waitaki River near Otematata and follows the river to the sea.

Travelling west from the Pacific Coast the landscape changes from green plains and rolling country to a harder, ruptured form and green gives way to brown and the grey of the underlying rock. Further inland the uplift of hills and ranges becomes more abrupt, and the course of the rivers more deeply etched, until it reaches its apogee in the lakes and mountain ranges of the Queenstown Lakes district. The western portion of the region is even more precipitous; Mt. Aspiring reaches 3036m. Many of the Western mountains are clad with thick, almost impassable bush, but some are so steep that they are virtually without any vegetation.

History

Whalers populated the fringe of the region in the 1830s, many of them intermarrying with the local Maori tribes. In Maori times Otago was fairly sparsely populated, one of the main attractions being the greenstone carried from the shores of Lake Wakatipu to the coast; in contrast, there is evidence that the area was quite densely settled in moa-hunter times.

The first significant influx of Europeans came at the end of the 1840s with the organised settlement established by the Lay Association of the Free Church of Scotland. From these arrivals has come much of the present character of the province. By the late 1850s runholders had extended the limits of settlements as far inland as Lake Wakatipu, but the greater proportion of population was concentrated near the coast and especially around Dunedin.

The sheep farmers were not destined to enjoy their isolation for long. In May 1861 Gabriel Read found a rich strike of gold on the Tuapeka River. This began the second, even greater inflow of settlers; by 1863 Otago had over 20,000 people, most of them risking death by drowning, exposure or starvation in the bleak central Otago winters in search of gold. The rushes subsided almost as quickly as they had begun, but the impetus given to the region's development lasted until the turn of the century. Although the miners moved off to Marlborough and the West Coast, the population continued to grow, especially in the 1870s when Otago received more assisted immigrants than any other part of New Zealand.

The prosperity brought by gold and wool reflected throughout the region. The plains and lowland hills were settled by arable farms and the larger holdings inland made some notable contributions to the development of agriculture. From the 1880s there was a revival of gold mining as dredges ground their way through river flats. By the turn of the century over 180 were operating, their courses marked by the arid tailings they left behind. They also left another legacy; many of the orchards found in central Otago were started in their wake.

The last gold boom had died by the 1920s, and Otago settled down to a steadier growth based on farming. Despite problems caused by rabbits and erosion, and a slower rate of development in recent decades, the region remains an important part of New Zealand economy.

Industry

New Zealand's frozen meat trade began from Port Chalmers when the **Dunedin** sailed in February 1882 with a load of mutton. Not surprisingly, sheep, and to a certain extent cattle, have remained an important source of Otago's prosperity. On the coastal area of north Otago stock farming is combined with mixed cropping; to the south dairying and fat lamb production is carried out. The higher and drier inland areas are used for store sheep and cattle raising.

In the dry harsh area of central Otago, stone-fruit orcharding has become a major industry. With a rainfall as low as 380 mm in some parts, this venture has only been made successful by the extensive use of irrigation. Many of the first irrigation schemes used old gold-miner's races and some still in service over 100 years later. In recent years even more ambitious schemes have begun; one on the lower Waitaki plain will raise the rainfall equivalent from 482mm to 1,016mm. By the end of the 1970s over 101,000 hectares should be under irrigation. Central Otago already produces 65 per cent of New Zealand's cherries, 60 per cent of nectarines, 60 per cent of plums and 90 per cent of apricots.

Forestry has become an important industry in the region. At Conical Hill is the largest timber mill in the South Island, and State exotic forests total over 20,000 hectares. Although present areas of exotic forests are relatively small compared to the North Island and even Nelson, planned expansion should allow the establishment of a wood pulp plant by the 1990s.

GAZETTEER OF NEW ZEALAND

The names are arranged alphabetically by islands

NORTH ISLAND

To find

Marsden Point **61**F3

This locality is shown on page **61** and within the square formed by the grid reference F3.

CITIES, TOWNS and LOCALITIES

Acacia Bay **69**H2
Admiral **75**E4
Admiral Hill **75**E4
Ahiaruhe **75**E3
Ahikiwi **60**C3
Ahikouka **74**D3
AHIPARA **58**C4
Ahirau **67**F5
Ahititi **68**D2
Ahurangi **73**E2
Ahuroa **62**B1
Akaaka **62**C5
Akaroa **75**G2
AKATARAWA **74**C3
Akatarawa Valley **74**C3
Akatere **59**E3
Akerama **61**E1
Akitio **75**H2
Akura **75**E3
ALBANY **62**C2
Albert Road **68**B4
Alfredton **75**F2
Alfriston **62**D3
Almadale **73**E4
Alton **72**C1
Anaura Bay **67**H4
Anawhata **62**B3
Aniwaniwa **70**D1
Aokautere **73**F5
Aomarama **60**E2
Aongatete **65**F1
Aorangi **67**G2
Aorangi **73**F5
Aoroa **60**D3
Aotea **64**C3
Apata **65**G2
Apiti **73**G4
Apiti Ridge **73**G4
Aponga **60**D2
Apotu **61**E2
Aputerewa **58**D4
Arahiwi **65**F4
Arakihi **67**G4
Aramatai **64**D5
Aramawhana **71**E2
Aramiro **64**C3
Aramoana **76**C4
Aranga **60**C2
Arapae **64**C5
Arapaoa **61**E4
Arapaoanui **70**D4
Araparera **62**B1
Arapata **73**F4
Arapohue **60**D3
Arapuni **65**E4
Ararata **68**C5
Ararewa **72**D2
Ararimu **62**D4
Ararua **61**E4
Arataki **65**F5
Aratapu **60**D4
Aratiatia **70**A1
Aratika **73**G5
Aratira **69**E5
Aratora **64**D5
Arawhata Road **68**A5
Ardkeen **71**E2
Ardmore **62**D3
Arero **67**H4
Argyll **76**B2
Argyll East **76**B2
Aria **69**E1
Arkles Bay **62**C2
Arohena **65**E4
Arowhana **67**E3
Ashcott **76**B3
ASHHURST **75**G5
Ashley Clinton **76**A3
Atea **75**E1
Atene **73**E2
Athenree **63**G5
Athol **65**F4
ATIAMURI **65**G5
AUCKLAND **62**C3
Aukopae **69**E3
Aupaki **76**C2
Auroa **68**B5
Avoca **60**D3
Awaawakino **67**E3
Awahou **65**G3
Awahou North **73**G5
Awahou South **73**G5
AWAHURI **73**F5
Awaiti **63**F5
Awakaponga **66**B2
AWAKERI **66**C3
Awakino **68**D1
Awakino Point **60**D3
Awamarino **64**B4
Awamate **71**E2
Awana **60**B4
AWANUI **58**C4
Awapiko **71**E1
Awapuni **71**G1
Awariki **73**H5
Awaroa **58**C5
Awarua **60**D2
Awatane **64**D4
Awatere Valley **67**H1
Awatoitoi **75**F3
Awatoto **70**C5
Awatuna **68**B5
Awhitu **62**C4
Awhitu Central **62**B4

Babylon **60**C3
Bainesse **73**E5
Bakers Crossing **73**E3
Balfour **67**E3
Ball Rd Rly Stn **72**C1
Ballance **75**F1
Barryville **65**E5
Bartletts **71**G2
Batley **61**E4
Battersea **74**D3
Baylys Beach **60**C3
BAY VIEW **70**C4
BEACHLANDS **62**D3
Beaconsfield **73**G4
BELL BLOCK **68**B3
Bells Junction **60**D2
Bells Junction **73**G2
BELMONT **74**B3
BELMONT HILL **74**B3
Belvedere **74**D3
Bennetts Siding **73**G2
Benneydale **64**D5
Bethells **62**B3
Bethlehem **65**G2
Bexhaven **67**G3
Bickerstaffe **61**E4
Bideford **75**F3
Big Bay **62**C3
Big Bay **63**F1
Big Hill **76**B1
Big Omaha **61**G5
Birch Hill **64**D4
BIRCHVILLE **74**C3
Blackburn **76**A2
Blackhead **76**C4
Black Stag **70**C3
Blairlogie **75**F3
BOMBAY **62**D4
Bonny Glen **73**E4
Bowentown **65**F1
Bradleys Landing **60**D3
Braemore **73**F3
Braigh **61**F4
Brancepeth **75**E3
BRIDGE PA **76**C1
Brighams Creek **62**B2
Brixton **68**B3
Broadlands **65**H5
Broadwood **58**D5
Brookby **62**D3
Brooklands **70**C5
Brooklands **68**A4
Brown Hill **74**D5
Brown Owl **74**C3
Brunswick **73**E3
Bruntwood **65**E2
Brynavon **61**F2
Brynderwyn **61**F4
BUCKLAND **62**D4
Bucklands **65**E3
BULLS **73**E4
BUNNYTHORPE **73**F5
Burney **75**F1
Burtenfields **75**F3
Bushgrove **75**F3
Bushy Knoll **71**E1
Bushy Range **76**B3

Cable Bay **58**D3
California **58**C4
CAMBRIDGE **65**E3
Canaan Landing **65**H2
Cape Colville **63**E1
Cape Runaway **67**G1
Cardiff **68**B4
Carnarvon **73**E5
Carrington **74**D3
Carswells Junction **75**F3
CARTERTON **75**E3
Castlehill **75**F2
Castlepoint **75**G3
Chases Gorge **60**C3
CHATEAU TONGARIRO **69**G4
Cheddar Valley **66**D3
Cheltenham **73**G4
Chudleigh Road **65**E1
Churchill **62**D5
Clareville **74**D3
Claris **60**B4
Clarks Beach **62**C4
Clevedon **62**D3
Cliff Rd Rly Stn **73**F4
Clifton **76**D1
CLIVE **76**C1
Cloustonville **74**C3
Cloverlea **73**F5
Clydebank **71**E3
Clydesdale **73**E5
Coal Creek **73**H4
Coatesville **62**C2
Colliers Junction **73**G2
Colliers Mill **65**E5
Colville **63**F2
Colyton **73**F5
Cooks Bay **63**G2
Coonoor **75**G1
Coopers Beach **58**D3
Cornwallis **62**C3
Coroglen **63**F3
COROMANDEL **63**F2
Cowes Bay **62**D3
Cricklewood **70**D2
Crofton **73**F4
Cross Creek **74**C3
Crownthorpe **76**B1
Cunninghams **73**F4

Dairy Flat **62**C2
Daisybank **74**D4
Dalefield **74**D3
DANNEVIRKE **73**H5
DARGAVILLE **60**D3
DAWSON FALLS **68**B4
Denlair **73**E3
Derby **68**B4
Devils Elbow **70**C4
Diggers Valley **58**C4
Dome **67**F4
Dome Valley **61**F5
Donnellys Crossing **60**C2
Douglas **68**C4
Dress Circle **73**H3
Dreyers Rock **75**F2
DRURY **62**D4
Duders Beach **62**D3
Dudley **68**B4
Dunmore **64**C2
Dunolly **73**F4
Durham **68**B4
Duthie Rd Rly Stn **68**B5
Dyerville **74**D4

EASTBOURNE **74**B4
East Cape **67**H1
East Clive **76**D1
East Tamaki **62**D3
East Taratahi **75**E3
Echolands **59**F2
EDGECUMBE **66**C2
Egmont Village **68**B3
Eight Mile Junction **64**C5
EKETAHUNA **75**E2
Elsthorpe **76**C3
Elstow **65**E1
ELTHAM **68**C5
Emerald Hills **71**F1
Endeans Mill **69**F1
Eparaima **75**F4
Erepiti **71**E1
Erewhon **73**H2
Erua **69**F4
Eskdale **70**C4
Eureka **65**E2
Everett **68**B3

Fairburn **58**D4
Fairfield **76**B3
Fairy Springs **65**G4
FEATHERSTON **74**C3
FEILDING **73**F5
Fencourt **65**E2
Fern Flat **58**D4
Fern Flats **73**E4
Fernglen **75**F4
FERNHILL **76**C1
Fernies **73**F3
Fernridge **75**E3
Fernside **67**G3
Fernside **74**D3
Fernyhurst **75**F3
Ferry Landing **63**F2
Flat Bush **62**D3
Flat Point **75**F4
Flemington **76**B4
Flock House **73**E4
Fordell **73**E3
Forest Gate **76**B3
Forest Lakes **74**D2
FOXTON **73**E5
Foxton Beach **73**E5
Frankley **68**B3
Fraser Road **68**C5
FRASERTOWN **71**E2
Frenchmans Swamp **59**G5
French Pass **65**E3
Front Beach **63**F2

GALATEA **66**B4
Galatea **73**E1
Gate **67**G2
GISBORNE **71**G1
Gladstone **74**D1
Gladstone **75**E3
Glen Afton **64**C1
Glen Avon **68**B3
Glen Bervie **61**E2
Glenbrook **62**C4
Glenbrook Beach **62**C4
Glenburn **75**E5
Glendhu **75**E5
Glenfalls **70**G3
Glengarry **73**H5
Glen Massey **64**C2
Glen Murray **62**D5
Glen Oroua **73**E5
Glenross **70**B5
Glenroy **67**G5
Glenshiel **75**E3
Glenside **74**B3
Glinkes Gully **60**D4
Glorit **62**B1
Golden Cross **63**G5
Golden Valley **63**G5
Gordon **65**F2
Gordonton **64**D2
Grahams Beach **62**C3
Greatford **73**F4
GREENHITHE **62**C2
GREYTOWN **74**D3
Gumtown **61**E2
Guthrie **65**G4
Gwavas **76**B2

Hahei **63**G3
Hairini **64**D3
Hakaru **61**F4
Haku **64**C5
HALCOMBE **73**F4
HAMILTON **64**D2
Hamua **75**F1
Hamurana **65**G3
Hangaroa **71**F1
Hangatiki **64**D4
Hannahs Bay **65**H4
Happy Jacks **71**G3
Happy Valley **62**D4
Harapepe **64**D3
Harrisville **62**D4
Haruru **59**F5
HASTINGS **76**C1
Hastwell **75**E2
Hatepe **69**H2
Hatfields Beach **62**C1
Hatuma **76**B3
HAUMOANA **76**D1
Hauiti **67**H4
Haunui **75**G2
Hauone **66**B2
Hauparu Bay **66**A3
Haupoto **67**E2
Haurangi **74**D4
Hautanoa **67**H3
Hautapu **65**E3
Hautere Cross **74**C2
Hautope **76**C3
Hautotara **74**D4
Hautu Prison **69**H3
Hauturu **64**C4
Hauturu **67**G2
HAVELOCK NTH **76**C2
Hawai **67**E2
HAWERA **68**C5
Hawkens Junction **66**B2
Hawkston **70**B4
Haywards **74**B3
Heao **69**E3
Heatherlea **73**D1
Heaton Park **73**E4
Heipipi **70**C1
Helena Bay **61**E1
HELENSVILLE **62**B2
Helvetia **62**C4
Henderson Valley **62**C3
Herbertville **76**B5
Hereheretau **71**F2
Herekino **58**C5
Hexton **67**G5
Hicks Bay **67**H1
Hihiroroa **67**F5
Hihitahi **73**G2
Hikawera **74**D4
Hikuai **63**G3
Hikumutu **69**F3
HIKURANGI **61**E2
Hikutaia **63**F4
Hillsborough **68**B3
Himatangi **73**E5
Himatangi Beach **73**E5
Hinakura **74**D4
Hinau **73**G3
Hinehopu **66**B3
Hinemoa **67**E4
Hinemoa **75**F2
Hinemoa Point **65**G4
Hinerua **76**A2
Hinuera **65**F3
Hiruharama **67**G2
Hiwera **73**H3
Hiwinui **73**F5
Hoanga **60**D3
HOBSONVILLE **62**C2
Hodderville **65**F4
Hoddle Corner **65**E4
Hoeotainui **65**E1
Hokio **73**D1
Hokoroa **67**G4
Holdens Bay **65**H4
Hollow Rock **74**D4
Homebrook **67**E4
Homeburn **74**C5
Homebush **75**E3
Homewood **75**F4
Honikiwi **64**C4
Hopelands **75**F1
Hopeone **66**D4
Hopuhopu **64**D2
Hopuruahine Landing **70**D1
Horahia **63**F4
Horahora **61**F2
Horahora **65**E3
Horeke **58**D5
Horoeka **67**G5
Horoeka **75**G1
Horoera **67**H1
Horohoro **65**G4
Horokiri Valley **74**B3
Horokiwi **74**B3
Horomanga **66**B4
Horopito **69**F4
HOROTIU **64**D2
Horsham Downs **64**D2
Hoteo **61**F5
Hoteo North **61**F5
Hot Water Beach **63**G3
Houhora **58**C3
Houto **60**D2
Huanui **67**G4
Huapai **62**B2
Huarau **61**E4
Huatokitoki **76**B4
Huia **62**B3
Huia **71**E1
Huiakama **68**C4
Huiarua **67**G3
Huinga **68**C4
Huirangi **68**B3
Huiroa **68**C4
Hukanui **67**G2
Hukanui **75**E1
Hukatere **58**C3
Hukatere **61**E4

MARAMARUA 63E4
Marangai 73E3
Marangairoa 67H1
Maratoto 63F5
Marco 68D3
Mareretu 61E4
Marima 75F1
Marlow 61E1
Maroa 65G5
Marohemo 61E4
Marokopa 64B4
Maromaku 61E1
Maropiu 60C3
Marotiri 69H1
Marotiri 74D1
Marsden Bay 61F3
MARSDEN POINT 61F3
MARTINBOROUGH 74D4
Martins Bay 62C1
MARTON 73F4
Marton Block 73H4
Marua 61E2
Marua 67F4
Marumaru 71E2
Marumoko 67E4
Maruroa 58C5
Marybank 73E3
Mason Ridge 76B2
MASSEY 62C3
MASTERTON 75E3
Mata 58D5
Mata 61E3
Mataatua 66C5
Matahanea 66D3
Matahapa 66D3
Matahi 66D4
Matahina 66B3
Matahiwi 76C1
Matahiwi 73E2
Matahiwi 75E3
Matahiwi Landing 73E2
Matahuru 63E5
Matai 65F2
Mataikona 75G3
Matakana 61G5
Matakatia Bay 62C2
Matakauwau 62C4
Matakohe 61E4
MATAMATA 65F2
Matamau 73H5
Matangi 64D2
Matangirau 59E4
Mataora Valley 63G5
Matapara 65E4
Matapihi 65G2
Matapiro 76B1
Matapouri 61F2
Matapouri 73G2
Matapu 68B5
Matarau 61E2
Mataraua 60C1
Mataraua 67G1
Matarawa 73E3
Matarawa 74D3
Mataroa 73G2
MATATA 66B2
Matatoki 63F4
Matau 68C4
Mataurangi 67F5
Matauri Bay 59F4
Matawai 67E4
Matawaia 60D1
Matawhera 60B1
Matawhero 71G1
Matawherohia 59E4
Matemateaonga 68C5
Matiere 69E2
Matingarahi 63E3
Matira 64C1
Matokitoki Valley 71G1
Matukuroa 72C1
Mauku 62C4
Maules Gorge 60C3
Maungahaumi 67F4
Maungakaramea 61E3
Maungaorangi 67G5
Maungapohatu 66C5
Maungarata 60D3
Maungaroa 67F1
MAUNGATAPERE 61E2
Maungatarata 67G3
Maungatautari 65E3
MAUNGATUROTO 61F4
Maunu 61E2
Mauriceville 75E2
Mauriceville West 75E2
Maxwell 72D2
Maxwell Rly Stn 72D2
Maymorn 74C3
MEEANEE 70C5
Mercer 62D4
MEREMERE 62D5
Meremere 68C5
Merita 58D3
MIDHURST 68B4
Mihi 65H5
Mihiwhetu 67G5
Mikimiki 75E2
Mill Creek 63F3
Mill Pa 71E2
Mimi 68C3
Mimiwhangata 61F1
Minden 65G2
Minginui 70C1
Miranda 63E4
Miromiro 70D2
Mitimiti 60B1
Mititai 60D4
Moanui 67E4
Moawhango 73H2
Moeatoa 64B5
Moeawatea 68D5
Moengawahine 60D2
Moerangi 64C3
Moerangi 69G2
MOEREWA 59F5
Moeroa 68D5
Moewhare 61E3
Mohaka 70D3
Mohakatino 68D2
Mohaka, Upper 70C3
Mohau 67G1
Mokai 65F5
Mokai 73H3
Mokara 70C3
Mokau 61E1
Mokau 68D1
Mokau River 68D2
Mokauiti 69E1
Mokoia 72C1
Mokopeka 76C2
Molesworth 61F4
Monavale 65E3
Moonshine 74C3
Morere 71F2
Morikau 73E1
Morland 75E3
MORRINSVILLE 65E2
Morrisons Bush 74D3
Mosston 73E3
Motairehe 60B3
Motakotako 64B3
Motatau 60D1
Motea 75H1
Moteo 70C5
Motu 67E4
Motuaruhe 67F1
Motuhora 67E4
Motuiti 73E5
Motukaraka 58D5
Motukauri 60B1
Motukiore 60C1
Motumaoho 65E2
Motunui 68B3
Motuoapa 69H3
Motutangi 58C3
Motuiti 58D5
Mototoa 60B1
Moumahaki 72D2
Moumoukai 63E4
Mount Biggs 73F4
Mount Bruce 75E2
Mount Curl 73F3
Mount Lees 73F4
MT MAUNGANUI 65G2
Mount Rex 62B2
Mount Richards 73G4
Mount Stewart 73F5
Mount Tiger 61F2
Mount Wesley 60D3
Mourea 65H3
Moutoa 75E1
Mowhanau 72D3
Muhunoa 74D1
Muhunoa East 74D1
Mullet Point 62C1
Murimotu 73G2
Muriwai 71G1
Muriwai Beach 62B2
Murumuru 69E4
MURUPARA 66B5
Muskers 68C3

Naike 64C1
NAPIER 70C5
Napinapi Pa 64C5
National Park 69F4
Naumai 60D4
Neavesville 63F4
Ness Valley 62D3
Netherby 64D1
NETHERTON 63F5
Newbury 73F5
Newman 75E2
NEW PLYMOUTH 68B3
Newstead 64D2
NGAERE 68C5
Ngahape 64D4
Ngahape 75F4
Ngahina 66C3
NGAHINAPOURI 64D3
Ngaiotonga 59G5
Ngaipo 75F4
Ngakaroa 67F5
Ngakonui 69F2
Ngakonui 75E4
Ngakuru 65G5
Ngamahanga 70D3
Ngamatapouri 72D1
Ngamatea 70A4
Ngamoko 73H4
Ngapaenga 64C5
Ngapaeruru 73H5
Ngapara 74D4
Ngapipito 60D1
Ngapotiki 74C5
Ngapuhi 60C1
Ngapuke 69F2
Ngapuna 65H4
Ngaputahi 66C5
Ngararatunua 61E2
Ngariki Road 69A4
Ngarimu Bay 63F4
Ngaroma 65E4
Nagroto 64D3
Ngaroto 70C5
Ngarua 65E2
NGARUAWAHIA 64D2
Ngataki 58C2
Ngatamahine 64D5
Ngatapa 67F5
Ngatarawa 76C1
NGATEA 63F4
Ngatimaru 68C4
Ngatimiro 68D3
Ngatira 65F4
Ngatiwhetu 58B2
Ngaturi 73E3
Ngaturi 75F1
Ngaumu 75E4
Ngaumu Forest Hqtrs 75E3
Ngaurukehu 73G2
Ngawaka 73G2
Ngawapurua 75F1
Ngawaro 65G3
Ngawha 59E5
NGAWHA SPRINGS 60D1
Ngawi 74C5
NGONGOTAHA 65G3
Ngongotaha Valley 65G4
Ngunguru 61F2
Ngutunui 64C3
Ngutuwera 72D2
Nihoniho 69E2
Nikau 75F1
Nireaha 75E1
NORMANBY 68C5
Norsewood 73H5
Norsewood South 76A3
North Egmont 68B4
NUHAKA 71F3
Nukuhau 70A1
Nukuhau 72D1
Nukuhou 66D3
Nukuhou North 66D3
Nukumaru 72D2
Nukutawhiti 60D2

Oakleigh 61E3
OAKURA 68A3
Oakura Bay 61E1
Oamaru Bay 63F2
Oaonui 68A5
Ocean Bch (Hastings) 76D2
Ocean Bch (N. Ak.) 61F3
Ocean Bch (Ohope) 66D3
Oeo 68A5
Ohaeawai 59F5
Ohakea 73E4
Ohaki 65H5
OHAKUNE 73G1
Ohakuri 65G5
Ohangai 72C1
Ohariu 74B3
Ohau 74D1
Ohaua Valley 66C3
Ohauiti 65G2
OHAUPO 64D3
Ohautira 64C2
Ohawe Beach 68B5
Ohinepaka 71E3
Ohinepanea 66B2
Ohinepoutea 67G2
Ohinewai 64D1
Ohinewairua 73H2
Ohingaiti 73G3
Ohiramoko 70D1
Ohirangi 62B2
Ohiwa 66D3
OHOPE 66C2
Ohotu 73H3
Ohui 63G4
Ohuka 71E2
OHURA 69E2
Ohuri 60C1
Oio 69F3
Okaeria 63E5
Okahu 60D4
Okahukura 69F2
OKAIAWA 68B5
OKAIHAU 59E5
Okaihau 67F4
Okaka 59E5
Okanagon 73E5
Okarae 73H5
Okare 71E2
Okareka 65H4
Okataina 66A3
OKATO 68A4
Okau 68D2
OKAUIA 65F2
Okauia Pa 65F2
Okawa 70C5
Okepuha 71G3
Okere Falls 65H3
Okete 64C2
Okiore 66D3
Okitu 71G1
Okiwi 60B4
Okoia 73E3
Okoki 68C3
OKOROIRE 65F3
Okoroire Rly Stn 65F3
Okotuku 72D2
Okupu 60B4
Okura 62C2
Okurakura 70C4
Omaha Flats 61G5
Omahaki 70B5
Omahanui 71E2
Omahina 72D2
Omahu 63F4
Omahu 76C1
Omahuta 58D5
Omaio 67E2
Omakere 76C3
Omakoi 70C1
Omamari 60C3
Omana 61E3
Omanaia 60C1
Omanawa 65G2
Omanawa Falls 65G2
Omapere 60B1
Omapere 76B1
Omaramutu 67E3
Omaranui 70C5
Omata 68A3
Omata 72D2
Omatane 73H3
Omaunu 59E4
Omeka Pa 65F3
Omiha 62D2
Omoana 68C5
Omokoroa 65G2
Omokoroa Beach 65G2
Omori 69G2
Onaero 68C3
Onekeneke Terraces 70A1
Onepu 66B3
Onepuhi 73F4
Onepunga 65E3
Oneriri 61E5
Oneroa 62D2
Onetangi 62D2
Onetea 60D3
One Tree Point 61F3
Onetohunga 67G3
ONEWHERO 62D5
Ongaha 74D4
Ongaonga 76B3
Ongare Point 65F1
Ongaroto 65F5
Ongarue 69F2
Oniao 69E1
Onini 66C5
Onoke 60B1
Opaea 73H2
Opaheke 62D4
Opahi 60D1
Opaki 75E3
Opaku 72C1
Opapa 76C2
Opape 67E3
Opara 60B1
Oparakau 61E4
Oparau 64C3
Oparure 64D4
Opatu 69E3
Opepe 70A2
Operiki Pa 73E2
OPIKI 75E1
Opoho 71F3
Opoiti 71E2
Oponae 66D4
Opononi 60B1
Oporae 75G1
OPOTIKI 66D3
Opouriao 66C3
Opoutama 71F3
Opouteke 60D2
Opoutere 63G4
Opua 59F5
Opua Road 68A5
Opuatia 62D5
Opuawhanga 61E1
OPUNAKE 68A5
Oraka 71G3
Orakau 64D3
Orakeikorako 65G5
Orangimea 72D2
Orangiponga 73G3
Oranoa 60C2
Oraora 60B2
Orapa 76B4
ORATIA 62C3
Orawau 58D5
Orautoha 69F4
Oreka 70C5
Oreore 63E3
Orere 63E3
Orere Point 63E3
OREWA 62C1
Oringi 73G5
Orini 64D1
Orira 58D5
ORMOND 67F5
Ormondville 76A3
Oromahoe 59F5
Orongo 63F4
Orongo Bay 59F5
Orongorongo 74B4
Oropi 75G2
Oroua Downs 73E5
Orton 62D5
Orua Bay 62C3
Oruaiti 59E4
Oruaiti Beach 67F1
Oruaiwi 69F2
Oruanui 69H1
Oruawharo 61F5
Orui 75F4
Oruru 58D4
Ostend 62D2
Otaha 59F4
Otahome 75F3
Otaihanga 74C2
Otaika 61E3
Otaika Valley 61E2
Otairi 73F3
Otakairangi 61E2
Otakehu 68B5
OTAKI 74C2
Otaki Beach 74C2
Otaki Forks 74D2
Otaki Gorge 74D2
Otakiri 66B3
Otamakapua 73G3
Otamarakau 66B2
Otamarakau Valley 66B2
Otamaroa 67G1
OTAMATEA 72D3
Otamauri 70B5
Otane 66D5
OTANE 76B2
Otangaroa 59E4
Otangiwai 69E2
Otara 66D3
Otaramarae 65H3
Otau 62D3
Otaua 60C1
Otaua 62C5
Otautu 63F1
Otautu 72C2
Otawhao 76A3
Otewa 64D4
Otiria 60D1
Otoi 70D2
Otoko 67E5
Otoko 73F2

BAYS, HARBOURS and LAKES

RIVERS and STREAMS

HEADLANDS and BEACHES

RANGES, MOUNTAINS and HILLS

HEADLANDS and BEACHES

ISLANDS

RANGES, MOUNTAINS and HILLS

METRIC MOTORING

METRIC MOTORING

New Zealand's decision to go metric is part of a world wide trend — ninety per cent of the world's population is already using the metric system or changing over to it.

To assist in the conversion from Imperial to Metric the charts below are included.

METRIC TERMS

The metric system is the simplest most readily understood method of measurement yet devised. Calculations are easy because the system builds on basic units in multiples of 10.

DISTANCE

All road signs and new map references now measure distance in **Kilometres** (km). One kilometre is very close to five-eighths of a mile. The kilometre is divided into **Metres** (m).

Miles	km	km	Miles
¼	0.40	1	0.62
½	0.80	2	1.24
¾	1.21	3	1.86
1	1.61	4	2.49
2	3.22	5	3.10
3	4.83	6	3.73
4	6.44	7	4.35
5	8.05	8	4.97
6	9.66	9	5.59
7	11.27	10	6.21
8	12.87	20	12.43
9	14.48	30	18.64
10	16.093	40	24.85
15	24.14	50	31.07
20	32.19	60	37.28
30	48.28	70	43.50
40	64.37	80	49.71
50	80.47	90	55.92
100	160.932	100	62.14
1000	1 609.32	1 000	621.40

Feet	m	Feet	m
1	0.305	20	6.10
2	0.610	30	9.14
3	0.914	40	12.19
4	1.22	50	15.24
5	1.52	60	18.29
6	1.83	70	21.34
7	2.13	80	24.38
8	2.44	90	27.43
9	2.74	100	30.48
10	3.05		

FUEL CONSUMPTION

The metric measure is litres per 100 kilometres (l/100 km). You will notice that this measurement is the reverse of miles-per-gallon, rather than a direct conversion. It is expressed in fuel units rather than distance. The resultant measure will be low for an economical car. Approximate equivalents are given below.
Miles per gallon (mpg)
Litres per 100 kilometres (l/100 km)

mpg	10	15	20	30	35	40
l/100 km	28	19	14	9	8	7

HEIGHT AND ELEVATION

Topographical elevation is shown on road signs and new maps in **metres** (m). One metre is about 3¼ ft. Height clearances for bridges and underpasses are also given in metres.

m	Feet	m	Feet
1	3.28	20	65.62
2	6.56	30	98.43
3	9.84	40	131.24
4	13.12	50	164.04
5	16.40	60	196.85
6	19.68	70	229.66
7	22.97	80	262.47
8	26.25	90	295.28
9	29.53	100	328.09
10	32.81		

BUYING PETROL AND OIL

The metric unit for fuel and oil is the **litre** (l). A litre is approximately 1¾ pints. Fuel and oil dispensers are now measured metrically.

l	gal.	l	gal.
1	0.22	20	4.40
2	0.44	30	6.60
3	0.66	40	8.80
4	0.88	50	11.00
5	1.10	60	13.20
6	1.32	70	15.40
7	1.54	80	17.60
8	1.76	90	19.80
9	1.98	100	22.00
10	2.20		

SPEED

From January 1st 1975 all vehicle speedometers must be calibrated in metric (km/h) readings. This conversion may be the fitting of a completely metric unit or a speedometer label clearly indicating the metric equivalents. Conversion of odo meters (mileage meters) is not required.

mph	km/h	mph	km/h
5	8.05	55	88.51
10	16.09	60	96.56
15	24.14	65	104.61
20	32.19	70	112.65
25	40.23	75	120.70
30	48.28	80	128.75
35	56.33	85	136.79
40	64.37	90	144.84
45	72.42	95	152.89
50	80.47	100	160.93

TYRE PRESSURE

The **kilopascal** (kPa) is the metric measurement for air pressure in tyres. One kilopascal equals .145 pounds per square inch. Alternatively 10 p.s.i. equals 69 kPa.

lbs.sq.in.	kPa	lbs.sq.in.	kPa
14	96.53	30	206.85
16	110.32	35	241.32
18	124.11	40	310.27
20	137.90	45	310.27
22	151.69	50	344.75
23	158.58	55	379.22
24	165.48	60	413.70
25	172.37	65	448.17
26	179.27	70	482.65
28	193.06	75	517.12

120

FIRST AID

LIFE SAVING MEASURES

AVOID BECOMING A CASUALTY YOURSELF — KNOW THE LIFE-SAVING PROCEDURES

1. The Situation

Be calm and take charge.

Ensure safety from traffic, fire and water, and the possibility of falling masonry, etc. Ask those present to remain if considered responsible as they may be able to help; otherwise they should be requested to stand clear.

Give each one a specific job, e.g.:

Ring up and notify the Police.

Ask for an Ambulance or,

Send for a Doctor.

In each case, state the place of the accident and tell what has happened.

Ask if anyone has any First Aid knowledge.

Ask for help in turning the casualty or, in steadying a limb.

In each case give exact instructions, and if necessary, show the bystander how your request should be carried out.

2. The casualty

Depending on what has happened, and the degree of severity of the injuries, and circumstances present, decided whether to treat the casualty where he is or whether to move him to a more suitable place. In a street accident note his exact position as the Police may want to know this.

Do not move the patient unless his life is in danger from some other cause.

If you decide to move him, carry out a quick preliminary examination of the head and neck, spine and four limbs. Then, decide on the most suitable method of removal in view of the injuries and the amount of skilled or unskilled help available. Then complete the examination of the casualty for injuries so that you can make a complete diagnosis and carry out the necessary treatment.

3. The disposal

Stay with the casualty and reassure him until Ambulance or Doctor arrives. Give your report to the Doctor and if necessary accompany the casualty to Hospital and report there. Notify the nearest relative and any other person or organisation that should be told. In serious outdoor accidents the Police should be sent for or notified.

The most urgent matters are:

1. To restart the heart if it has stopped beating.
2. To apply Resuscitation if the casualty is not breathing. If in doubt as to whether the casualty is alive or not, continue treatment until medical aid is available.
3. To control bleeding.
4. To maintain a clear airway by correctly positioning the casualty.

The most important procedures to prevent the condition becoming worse are:

1. To dress wounds.
2. To immobilise fractures and large wounds.
3. To place the casualty in the most comfortable position, consistent with the requirements of treatment.

The most helpful measures in promoting recovery are:

1. To relieve the casualty of anxiety and promote his confidence.
2. To relieve him of pain and discomfort.
3. To protect him from the cold.
4. To handle gently so as to do no harm.

OBSTRUCTION OF THE AIRWAY

If an unconscious person is lying on his back there is the danger of the tongue falling back and blocking the airway. There is also the danger of secretions or regurgitated stomach contents entering the windpipe because of the epiglottis failing to function, which is why the coma (semi-prone) or recovery position of casualty must be used.

BREATHING

Breathing, it goes without saying, is necessary for the maintenance of life and this must be assured in every case of accident or sudden illness. The conscious patient is usually able to maintain an open airway for himself and so does not present any great problem, but the unconscious patient is in a vastly different state and requires constant supervision to ensure that he is able to breathe. The tissues in the throat become flaccid and sag, the tongue falls back and obstructs, or may completely block the air way, and if this occurs then death will quickly follow. No matter what position the patient's body is in if his head is tilted back into the "sniffing" position then the tissues of the throat will be tightened and the base of the tongue lifted off the top of the breathing passages. This will enable the patient to breathe for himself if he is able. But if the lifting of the chin and extending the head does not result in spontaneous breathing then resuscitation must be instituted at once. This is quite simply done by pinching the nostrils and breathing into the patient's mouth to inflate the lungs.

In many cases only a few breaths will be necessary, but after the first six an attempt should be made to find a heartbeat. This is done at the large artery in the neck (carotid). If no pulse is found a heartbeat must be provided for the patient by what is known as Closed Chest Cardiac Compression, as without a circulation of blood brain cells will die and the condition become irreversible within four minutes.

This simple life saving technique can be easily learned by anyone, young or old, and if you have not already done so, then you should attend a demonstration as soon as possible.

WOUNDS WITH SEVERE BLEEDING

1 Place the casualty at absolute rest, preferably with the legs raised.
2 Elevate the injured part unless an underlying fracture is suspected.
3 Expose the wound removing as little clothing as possible.
4 Do not waste time washing your hands or cleaning the wound area in case of severe haemorrhage.
5 Control haemorrhage by grasping sides of wound firmly together; if more expedient, by direct digital pressure on the bleeding point or points, preferably over a clean dressing. Avoid disturbing blood clots.
6 Apply sufficient sterile dressing packed into the depth of the wound, until it projects above the wound, then cover with adequate padding and bandage firmly.
7 If bleeding continues, do not disturb the original dressing or bandage, but, add additional pads and again bandage firmly.
8 a) If foreign bodies are present in the wound and are visible and capable of removal, gently wipe off with a clean dressing, otherwise leave alone and proceed as in paragraph (b)
  b) If a foreign body has to be left in the wound, or bone is projecting, cover the wound with a sterile dressing, and apply pads round the wound to a sufficient height to enable the bandage to be applied in a diagonal manner avoiding pressure on the projecting foreign body or bone.
9. All bandages should be applied just sufficiently firmly to stop bleeding.
10 Immobilise the injured part by a suitable method — e.g. a sling in the case of the upper limb, or by tying an injured lower limb to the uninjured one.
11 Keep casualty comfortably warm with blankets
12 Remove him to hospital as quickly as possible.

BURNS AND SCALDS

General rules for the treatment of burns and scalds.

1 No matter which part of the body is affected, immerse in cold water, or place part gently under cold running water. **AT ONCE**
2 Do not remove burnt clothing as it will have been rendered sterile by heat.
3 Do not break blisters and keep immersed in cold water if still painful.
4 Remove at once anything of a constricting nature — rings, bangles, belts, boots — before the part starts to swell.
5 Cover the area with a sterile dressing, clean lint or freshly laundered linen.
6 If liable to get dirty, as in the case of a hand or foot, apply sterile or clean smooth dressing lightly.
7 If the area burned is **larger** than the **palm of the patients hand** get him to medical aid quickly.

INJURIES TO BONES

Fractures

A "fracture" is a broken or cracked bone. Where there is a history of force applied to a bone and the diagnosis is uncertain treat all such injuries as fractures.

Type of fracture

1 CLOSED OR SIMPLE: When there is no wound leading down to the broken bone.
2 OPEN OR COMPOUND: When there is a wound leading down to the broken bone, or when the fractured ends protrude through the skin thus allowing germs to gain access to the site of the fracture.
3 COMPLICATED: When there is some other injury directly associated with the fracture, such as to an important blood vessel, the brain, nerves, lungs or when associates with a dislocation.

General signs and symptoms of fractures

Comparison with the uninjured side with often help in diagnosis .
1 Pain over the injured part.
2 Tenderness on gentle pressure
3 Swelling and later bruising
4 Loss of control
5 Deformity of the limb, such as shortening or angularity (bend appearing in unusual position).
6 Irregularity of the bone may be felt.

General rules for the treatment of fractures

1 Severe wounds and bleeding must be dealt with before continuing with treatment of fractures.
2 Treat the fracture where the casualty lies. The injured part must be secured, even if only in a temporary way before the casualty is moved, unless life is immediately endangered.
3 Steady and support the injured part at once, and maintain this control until such time as the fracture is completely secured.
4 Immobilise the fracture either by securing the injured part to a sound part of the body by means of bandages or, where necessary, by the use of splints and bandages.

Using bandages

Bandages should be applied sufficiently firmly to prevent movement, but not so tightly as to prevent the circulation of the blood.

If the casualty is lying down, use a splint or similar object to pass the bandage under the truck or lower limbs in the natural hollows of the neck, waist, knees and just above the heels. The bandages may then be worked gently into their correct position.

Priorities

If you are clear about the priorities in an accident (that is, the order in which things are done) you are well on the way to being a First Aider. The following are the priorities:

Danger

If you do not deal with threatened danger you and your casualties may be killed. PILE-UPS and FIRE are the dangers in a road accident.
Action
Get someone to flag down the traffic far enough away to secure compliance.
Switch of the engine.
Impose a 'No Smoking' ban

Breathing

A crash victim is often unconscious and cannot breathe because of a kink in his airway.
Action
Open the airway up by extending the head backwards.
Check for obstruction to airway and relieve it.
If still not breathing pinch nose, hold head back and inflate lungs by blowing.

Bleeding

Grasp sides of wound
Elevate if possible
Continue pressure on sides of wound with pad and firm bandage.

Coma

You must keep the unconscious patient alive until the ambulance arrives:
Action
Keep the airway open
Turn into the coma or recovery position
Watch the airway
WATCH THE AIRWAY

Shock

In severe injuries the patient will die in a few hours unless he gets a blood transfusion.
Action
Send for the ambulance with extreme urgency.

Fractures

Immobilise using common sense.
Upper limb: use arm sling or pin sleeve to lapel
Lower limb: tie to sound leg after padding between knees and ankles.

Wounds

Stop bleeding
Clean around wound with throw-away tissues
Cover with sterile or clean dressing
Immobilise.

Dial 111 and ask for the ambulance service, stating:
Where to come
How many patients.
The nature of the injuries

AA SERVICES

THE AUTOMOBILE ASSOCIATION AND YOU

Consider the motor car. Somewhere in our history Man's inventiveness may have come up with more significant, more astonishing developments than this, but surely none can match the motor car in so completely satisfying a human need. For we are travelling creatures, always on the go in search of goods, work, companionship, or recreation. The motor car was a revolution in personal transport.

There are those who say that the car has reached a dead end, a crisis stage of urban congestion, polluted air, desecrated environment, and an intolerable road toll. They want to restrict the quantity and mobility of the car, hamper its usefulness, perhaps even roll back the years of progress to eliminate it altogether.

But the people want and need the motor car more than ever, and every year brings improvements in its design, function, and safety. Since the turn of the century its popularity has grown like a whirlwind, and our lives and horizons have been enriched beyond the imagination of the earliest motorists. Today the motor vehicle is woven into the pattern of living.

It is not surprising, then, that this relationship gives rise to a truly unique and universal movement — the world's Automobile Associations. Founded in each country by keen motoring pioneers who banded together in the early years, the AA grew in membership almost as fast as cars were manufactured. Today this movement ignores political boundaries to include over 130 national organisations and 30 million members.

In New Zealand the AA has been particularly active right from the beginning. Less than two years after the first car appeared in Auckland in 1900, the first AA was founded and the battle joined to gain recognition for this new horseless carriage. It seems ironic that, more than 70 years later, the AA once again is defending the motor car against short-sighted criticism.

The transition from motoring club, for the sharing of travel pleasures and improvement of motoring conditions, to wide-ranging service organisation developed during the 1920s. First to appear were road maps, the backbone of AA service throughout the world. Then accommodation handbooks and touring assistance, technical advice, and road signs — New Zealand is one of the very few countries in the world where the AA, not the Government, is entrusted with signposting.

But throughout this development the AA has never lost sight of its original aim, the continuing effort to achieve better roads, safer traffic, more pleasant holidays, and trouble-free motoring. And because the AA has earned a reputation for integrity and responsibility, these efforts are often successful. Perhaps, after all, it's because motorists are the people, and the people are the nation.

Today the New Zealand Automobile Association links 15 separate and autonomous associations. Each provides services for its own area, and each has its own set of rules and range of services. Yet any AA member, whether from another district or another country, is welcome to the same service afforded a local member.

There is an AA office in every city and large town, and most smaller towns have agencies dispensing touring advice and maps. In the main centres, the scope of services available is remarkable — expert car inspections, on-the-spot breakdown assistance, legal aid for traffic offences, travel and accommodation booking, hire purchase loans, car insurance, driving instruction, and car shipping assistance are just some of the benefits.

New Zealand is an ideal country to tour by motor car, which may explain why vehicle density per population here ranks third in the world. City or country, cars are vital to duty fulfilment and family enjoyment. The AA is at your service, on behalf of half a million New Zealand motorists, to get the most out of your motoring.

GLOSSARY OF MAORI WORDS

Common Maori words found in simple or compound form in place names throughout New Zealand: —

Nouns

awa	— river, creek	rau	— multitude, hundred
kai	— food	repo	— swamp
kaingi	— village	roto	— lake
kura	— crayfish	rua	— hole, hollow, cave
manu	— bird	tane	— man
maunga	— mountain	tapu	— sacred
moana	— lake, sea	wahine	— woman
moko	— tattoo	wai	— water
motu	— island	waka	— canoe
one	— mud, sand, beach	whanga	— bay
pa	— fortified village	whare	— house
pohatu	— stone	whata	— raised platform for storing food
rangi	— sky	whenua	— land, country

Adjectives

ehu	— muddy	nui	— big
heke	— falling	pai	— good
iti	— small	roa	— long, tall
kawa	— bitter	tapu	— sacred
kino	— bad	tere	— fast
makariri	— cold	wera	— hot
ngaro	— lost		

Articles

te — singular
nga — plural } the

Numerals

rau — hundred
tahi — one
rua — two

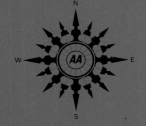